DATE DUE

FINE CHEMICALS

FINE CHEMICALS
THE INDUSTRY AND THE BUSINESS

SECOND EDITION

Peter Pollak, PhD
Reinach, Switzerland

A JOHN WILEY & SONS, INC., PUBLICATION

Published by John Wiley & Sons, Inc., Hoboken, New Jersey.
Published simultaneously in Canada.

For general information on our other products and services or for technical support, please contact our Customer Care Department within the United States at (800) 762-2974, outside the United States at (317) 572-3993 or fax (317) 572-4002.

Wiley also publishes its books in a variety of electronic formats. Some content that appears in print may not be available in electronic books. For more information about Wiley products, visit our web site at www.wiley.com.

Library of Congress Cataloging-in-Publication Data:
Fine chemicals : the industry and the business/edited by Peter Pollak.—2nd ed.
    p. cm.
    ISBN 978-0-470-62767-9 (cloth)
    1. Chemicals. 2. Chemical engineering. 3. Chemical industry. I. Pollak, Peter, 1934–
    TP200.F525 2010
    660—dc22
                                                    2010033575

To Maria, Barbara, and Paolo

CONTENTS

NOTE:
US$ Exchange rates

| CHF 1 = US$ 0.966 | € 1 = US$ 1.44 | INR 100 = US$ 2.15 | £ 1 = US$ 1.52 |

■ PREFACE TO THE SECOND EDITION

This revised edition has been prepared in order to provide the reader with an updated view of the fine chemical industry and business. Actually, the most recent data used in the first edition go up to 2005. Since then, both the fine chemicals industry itself and its customers have undergone substantial changes: The business as a whole is more competitive now due to the escalating impact of low-cost players from the Far East on the one hand, and the passing-on of the price pressure the life science industry is subject to itself on the other hand. Many Western fine chemical companies or divisions created during the "irrational exuberance" at the turn of the millennium have exited from the sector. As described in Section 5.2, Plant Operation, the most progressive companies adopt lean production principles originally developed by the automotive industry. In this context, Section 7.2, Manufacturing Costs, has also been refined. As described in the new Section 2.5, Mergers and Acquisitions, the M&A scenario has inverted from a seller's to a buyer's market. Private equity firms have become owners of a number of fine chemical companies. The business model of the fine chemical industry has broadened and now often includes also contract research at the beginning and active ingredient formulation toward the end of the value-added chain. This development is not finding unanimous approval by industry experts. Biotechnology now plays a bigger role. In the synthesis of small molecules, the use of enzymes has become more widespread as it enables both a more sustainable and economic production. Big molecules have firmly established themselves as active ingredients in the life science industry. Thus, biopharmaceuticals now account for five of the ten top selling pharmaceuticals. To the detriment of originator drugs and agrochemicals, the market share of generics has increased. It now also comprises generic versions of biopharmaceuticals (see the new Section 17.2, Biosimilars).

Faced with slower growth, patent expirations of many lucrative blockbuster drugs, and stalling new product launches, leading pharmaceutical companies are facing challenges as never before. They react by implementing restructuring programs. These comprise, among others, reduction of their in-house chemical manufacturing and plant eliminations. Outsourcing of chemical manufacturing has moved up from a purely opportunistic to a strategic approach. Apart from restructuring, globalization is also affecting the fine chemicals industry (see Chapter 14). In the pharm-emerging countries, double-digit growth of the consumption of pharmaceuticals and agrochemicals—*and* the production of their active ingredients—is taking place. With their combination of 40% of the world's population and the "low-cost/high-skill/high-future" industrial base, they represent a great challenge to the European and U.S. fine chemicals industry. A business condition could develop, whereby even "best-in-class" midsized, family-owned fine chemical companies with superior technology portfolios and footholds in Asia could be relegated to producing small quantities of fine chemicals for new life science products in late stages of development (see Chapter 20).

PETER POLLAK

■ PREFACE TO THE FIRST EDITION

This book provides an insider's view of the status of the fine chemical industry, as well as its outlook. It covers all aspects of this dynamic industry, with all of its stakeholders in mind, viz. employees, customers, suppliers, investors, students and educators, media representatives, neighboring communities, public officials, and anyone else who has an interest in this segment of the chemical industry. Safety, health, environmental, and regulatory issues are discussed only briefly, as the related subjects are extensively covered in the specialized literature.

The main raison d'être of the fine chemical industry is to satisfy the product and process development needs of the specialty chemicals, especially the life science (primarily pharmaceutical and agrochemical) industry. Sales outside the chemical industry remain the exception. The fine chemical industry has evolved mainly because of the rapid growth of the Anglo-Saxon pharmaceutical industry, which traditionally has been more inclined to outsourcing chemical manufacturing than the continental European one—and the increasing complexity of the drug molecules. The roots of both the term "fine chemicals" and the emergence of the industry as a distinct entity date back to the late 1970s, when the overwhelming success of the histamine H_2 receptor antagonists Tagamet (cimetidine) and Zantac (ranitidine hydrochloride) created a strong demand for advanced intermediates used in their manufacturing processes. The two drugs cure stomach ulcers, thus eliminating the need for surgical removal of ulcers. As the in-house production capacities of the originators, Smith, Kline & French and Glaxo, could not keep pace with the rapidly increasing requirements, both companies outsourced part of the synthesis to chemical companies in Europe and Japan experienced in producing relatively sophisticated organic molecules. Also, the fledgling generics industry had no captive production of active pharmaceutical ingredients (APIs) and purchased their requirements. Moreover, the growing complexity of pharmaceutical and agrochemical molecules and the advent of biopharmaceuticals had a major impact on the evolution of the fine chemical industry as a distinct entity. Custom manufacturing, respectively its counterpart, outsourcing, has remained the *Königsdisziplin* (i.e., the most prominent activity) of the fine chemical industry and "make or buy" decisions have become an integral part of the supply chain management process. The fine chemical industry has its own characteristics with regard to R&D, production, marketing, and finance. The total turnover of the largest companies, respectively business units does not exceed a few hundred million dollars per year. The fine chemical industry supplies advanced intermediates and active substances, frequently on an exclusive basis, to the pharmaceutical, agrochemical, and other specialty-chemical industries. Further distinctions are batch production in campaigns, high asset intensity, and above-industry-average R&D expenditures. The industry is still located primarily in Europe. Custom manufacturing prevails in northern Europe; the manufacture of active substances for generics, in southern Europe.

As of today, the majority of the global $85 billion production value of fine chemicals continues to be covered by captive production, leaving a business *potential* of $60 billion for the fine chemical industry ... on top of the inherent growth of the existing business. Despite this huge business opportunity, the fine chemical industry is challenged by overcapacity and intense competition. As a result of early riches, many chemical companies sought relief from their dependence on cyclical commodities by diversifying into higher-value-added products, like fine chemicals. At present, the industry is going through two interconnected changes. In terms of geography, Far Eastern "high-skill/low-cost" companies are emerging as serious competitors. In terms of structure, the chemical conglomerates are divesting their (often loss-making) fine chemical businesses. They are becoming mostly privately owned pure players. Although the demand has not grown to the extent initially anticipated, fine chemicals still provide attractive opportunities to well-run companies, which are fostering the critical success factors, namely running fine chemicals as core business, making niche technologies—primarily biotechnology—a part of their business and developing assets in Asia.

PETER POLLAK

ACKNOWLEDGMENTS

SECOND EDITION

For the preparation of the second edition of this book, I am indebted to a number of colleagues, who, as experts in specific fields, provided me with their very valuable inputs: Rolf Dach, Boehringer Ingelheim; Jacques Gosteli, Cerecon AG; Vihari Purushothaman, Enam Securities Pvt Ltd.; Wilfried Eul and Dan Ostgard, Evonik Degussa; Anish Swadi, Hikal Ltd, Stefan Stoffel and Robert Voeffray, Lonza Ltd.; Ian Shott, Shott Consulting, Andrew Warmington, *Speciality Chemicals Magazine*, and Dimitrios Kalias, Vio Chemicals Ltd.

FIRST EDITION

I would like to acknowledge all individuals, both peers and customers from my present consulting activity, and colleagues from my former association with Lonza, who have helped me in conceiving, writing, and reviewing this book. I am particularly indebted to Rob Bryant (Brychem) and Ian Shott (Excelsyn), who have shared with me both their profound knowledge of and their ability to communicate with the industry. I am also very grateful for the valuable input, whether in providing data or in proofreading, that the following individuals have kindly provided: Vittorio Bozzoli, Ron Brandt, Uli Daum, Peter Demcho, Erich Habegger, Wouter Huizinga, Mario Jaeckel, Myung-Chol Kang, Dr. Masao Kato, Christine Menz, Hans Noetzli, H. Barry Robins, and Carlos Rosas.

Without this invaluable assistance from these friends and colleagues, I would not have been able to embark on this ambitious undertaking.

PART I

THE INDUSTRY

CHAPTER 1

What Fine Chemicals Are

1.1 DEFINITION

The underlying principle for definition of the term "fine chemicals" is a three-tier segmentation of the universe of chemicals into commodities, fine chemicals, and specialty chemicals (see Fig. 1.1). Fine chemicals account for the smallest part, about 4% of the total $2500 billion turnover of the global chemical industry (see Section 9.1).

Commodities are large-volume, low-price homogeneous, and standardized chemicals produced in dedicated plants and used for a large variety of applications. Prices, typically less than $1/kg, are cyclic and are fully transparent. Petrochemicals, basic chemicals, heavy organic and inorganic chemicals, (large-volume) monomers, commodity fibers, and plastics are all part of commodities. Typical examples of single products are ethylene, propylene, acrylonitrile, caprolactam, methanol, toluene, *o*-xylene, phthalic anhydride, poly (vinyl chloride) soda, and sulfuric acid.

Fine chemicals are complex, single, pure chemical substances. They are produced in limited quantities (up to 1000 MT per year) in multipurpose plants by multistep batch chemical or biotech(nological) processes. They are based on exacting specifications, are used for further processing within the chemical industry, and are sold for more than $10/kg (see Fig. 1.1).

> Fine chemicals are "high-value chemicals purchased for their molecular qualities rather than for their functional performance, usually to make drugs."
>
> –Rick Mullin, *Chemical & Engineering News*

The category is further subdivided on the basis of either the added value (building blocks, advanced intermediates, or active ingredients) or the type of business transaction (standard or exclusive products). As the term indicates, *exclusive products* are made exclusively by one manufacturer for one customer, which typically uses them for the manufacture of a patented specialty

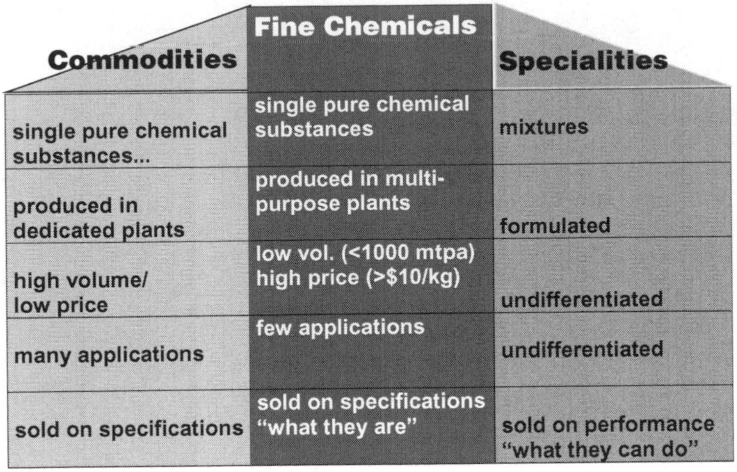

Commodities	Fine Chemicals	Specialities
single pure chemical substances...	single pure chemical substances	mixtures
produced in dedicated plants	produced in multi-purpose plants	formulated
high volume/ low price	low vol. (<1000 mtpa) high price (>$10/kg)	undifferentiated
many applications	few applications	undifferentiated
sold on specifications	sold on specifications "what they are"	sold on performance "what they can do"

Figure 1.1   Definitions.

chemical, primarily a drug or agrochemical. Typical examples of single products are β-lactams, imidazoles, pyrazoles, triazoles, tetrazoles, pyridine, pyrimidines, and other *N*-heterocyclic compounds (see Section 3.1). A third way of differentiation is the regulatory status, which governs the manufacture. Active pharmaceutical ingredients (APIs) and advanced intermediates thereof have to be produced under current Good Manufacturing Practice (cGMP) regulations. They are established by the (U.S.) Food and Drug Administration (FDA) in order to guarantee the highest possible safety of the drugs made thereof. All advanced intermediates and APIs destined for drugs and other specialty chemicals destined for human consumption on the U.S. market have to be produced according to cGMP rules, regardless of the location of the plant. The regulations apply to all manufacturing processes, such as chemical synthesis, biotechnology, extraction, and recovery from natural sources. All in all, the majority of fine chemicals have to be manufactured according to the cGMP regime.

A precise distinction between *commodities* and *fine chemicals* is not feasible. In very broad terms, commodities are made by chemical engineers and fine chemicals by chemists. Both commodities and fine chemicals are identified according to specifications. Both are sold within the chemical industry, and customers know how to use them better than do suppliers. In terms of volume, the dividing line comes at about 1000 tons/year; in terms of unit sales prices, this is set at about $10/kg. Both numbers are somewhat arbitrary and controversial. Many large chemical companies include larger-volume/lower-unit-price products, so they can claim to have a large fine chemicals business (which is more appealing than commodities!). The threshold numbers also cut sometimes right into otherwise consistent product groups. This is, for instance, the case for

APIs, amino acids, and vitamins. In all three cases, the two largest-volume products, namely, acetyl salicylic acid and paracetamol; L-lysine and D,L-methionine, and ascorbic acid and niacin, respectively, are produced in quantities exceeding 10,000 tons/year, and are sold at prices below the $10/kg level.

Specialty chemicals are formulations of chemicals containing one or more fine chemicals as active ingredients. They are identified according to performance properties. Customers are mostly trades outside the chemical industry and the public. Specialty chemicals are usually sold under brand names. Suppliers have to provide product information. Subcategories are adhesives, agrochemicals, biocides, catalysts, dyestuffs and pigments, enzymes, electronic chemicals, flavors and fragrances, food and feed additives, pharmaceuticals, and specialty polymers (see Chapter 11).

The distinction between fine and specialty chemicals is net. The former are sold on the basis of "what they are"; the latter, on "what they can do." In the life science industry, the active ingredients of drugs, also known as APIs or drug substances (DS), are fine chemicals, the formulated drugs specialties, aka drug products (DP) (see Chapter 2).

Electronic chemicals (see Section 11.4.5) provide another illustrative example of the difference between fine and specialty chemicals: Merck KGaA produces a range of individual fine chemicals as active substances for liquid crystals in a modern multipurpose plant in Darmstadt, Germany. An example is (*trans,trans*)-4-[difluoromethoxy)-3,5-difluorophenyl]-4′-propyl-1,1′-bicyclohexyl. Merck ships the active ingredients to its secondary plants in Japan, South Korea, and Taiwan, where they are compounded into liquid crystal formulations. These specialties have to comply with stringent use-related specifications (electrical and color properties, etc.) of the Asian producers of consumer electronics such as cellular phones, DVD players, and flat-screen TV sets.

"Commoditized" specialty chemicals contain commodities as active ingredients and are interchangeable. Thus, ethylene glycol "99%" is a commodity. If it is diluted with water, enhanced with a colorant, and sold as "super antifreeze" in a retail shop, it becomes a commoditized specialty.

Note: Sometimes fine chemicals are considered as a subcategory of specialty chemicals. On the basis of the definitions given above, this classification should be avoided.

1.2 POSITIONING ON THE VALUE-ADDED CHAIN

An example of the value-added chain extending from commodities through fine chemicals to a pharmaceutical specialty is shown in Table 1.1. The product chosen is Pfizer's anticholesterol drug Lipitor (atorvastatin), the world's top-selling drug with sales of $11.4 billion in 2009.

The value-added chain extends from a C_1 molecule, methanol (left side of the table), all the way to a C_{33} molecule, atorvastatin. The indicative $2000 cost

TABLE 1.1   Example for the Value-Added Chain in the Chemical Industry: Lipitor (Atorvastatin)

Parameter	Commodities		Fine Chemicals			Specialty
			Intermediates		API	
Example	Methanol	Acetic Acid	(I)	(II)	(III)	Lipitor
Molecular formula	CH_4O	$C_2H_4O_2$	$C_7H_{11}NO_3$	$C_{14}H_{30}NO_4$	$C_{33}H_{35}FN_2O_5$	N/A
Molecular weight	32.04	60.05	157.17	269.34	558.65	N/A
Applications	>100	>50	10	1	1	N/A
Price indication ($/kg)	0.2	1.00	100	200	2000	80,000
Production (metric tons/ year)	32×10^6	8×10^6	200	300	400	400
Producers	100	25	10	5	1^a	1^a
Customers	100	50	1	1	1^a	Millions
Plant type[b]	D, C	D, C	M, B	M, B	M, B	F&P
Manufacturing steps	1	2	5	15	20	+2

[a] Active pharmaceutical ingredient.
[b] Patent holder; several generic producers preparing for launch.

Note: Figures are indicative only.

Key: B, batch; C, continuous; D, dedicated; M, multipurpose.

(I) Ethyl (*R*)-4-cyano-3-hydroxy butanoate, "hydroxynitrile."
(II) tert-Butyl (4*R*,6*R*)-2-[6-(2-aminoethyl)-2,2-dimethyl-1,3-dioxan-4-yl]acetate.
(III) 2-(4-Fluorophenyl)-β,δ-dihydroxy-5-(1-methylethyl)-3-phenyl-4-[(phenylamino)-carbonyl]-1H-pyrrole-heptanoic acid.

per kilogram API corresponds to about 4% of the price of the formulated prescription drug ($400 for 100 80mg tablets) bought in the pharmacy. The structure of compound **III** in Table 1.1 is as follows:

(III)

Methanol and acetic acid are typical commodities, namely, low-price/polyvalent products manufactured in large quantities by many companies. Under the heading "fine chemicals," three molecules used for the manufacture of atorvastatin are listed, namely, the advanced intermediates ethyl 4-chloro-3-hydroxy butanoate (**I**) and *tert*-butyl (4*R*,6*R*)-2-[6-(2-aminoethyl)-2.2-dimethyl-1.3-dioxan-4-yl] acetate (**II**), respectively, and the API, atorvastatin (**III**) itself. As long as the latter, 2-(4-fluorophenyl)-β,δ-dihydroxy-5-(1-methylethyl)-3-phenyl-4-[(phenylamino)-carbonyl]-1*H*-pyrrole-heptanoic acid, is sold according to specifications, it is a fine chemical. In the pharmaceutical industry, the chemical synthesis of an API is also referred to as *primary manufacturing*. The secondary manufacturing comprises the formulation of the API into the final delivery form. The API is compounded with excipients that confer bulkiness, stability, color, and taste. Once atorvastatin is tableted, packed, and sold as the anticholesterol prescription drug Lipitor, it becomes a specialty.

CHAPTER 2

The Fine Chemical Industry

2.1 INDUSTRY STRUCTURE

Within the chemical universe, the fine chemical industry is positioned between the commodity and specialty chemical industries. The latter are their suppliers and customers, respectively. Among the customers, life sciences, especially the pharmaceutical industry, prevail (see Section 9.2). In the broad sense, fine chemical companies are active in research and development, manufacturing, and marketing of fine chemicals. They represent a wide variety of several 1000 enterprises offering mainly products and services along the drug supply chain (see Fig. 2.1). They extend from small, privately owned laboratories all the way to large, publicly owned manufacturing companies. However, not all of them encompass all three activities. Whereas Western fine chemical companies still dominate in sales revenues, most of the small ones are located in Asia.

Fine chemical/custom manufacturing (CM) companies in the narrower sense are active in process scale-up, pilot plant (trial) production, and industrial-scale exclusive and nonexclusive manufacture and marketing. Their product portfolios comprise exclusive products, produced by CM (see Section 12.2.1), nonexclusive products, for example, active pharmaceutical ingredient (API)-for-generics (12.2.2), and standard products (12.2.3).

Contract research organizations (CROs, see Section 2.3) provide process development and bench scale synthesis services. Fine chemical companies which are both CM companies and CROs are called *contract research and manufacturing companies* (CRAMS).

Laboratory chemical suppliers (see Section 2.4) offer a large number (thousands) of different kinds of chemicals in small quantities for research purposes.

Finally, there are firms that do neither contract research nor manufacturing. They are *distributors* and/or *agents* of integrated fine chemical companies.

Fine chemical/CM companies account for the largest share of the industry, followed by chemical CROs and laboratory chemical suppliers.

Fine Chemicals: The Industry and the Business, Second Edition, by Peter Pollak
Copyright © 2011 John Wiley & Sons, Inc.

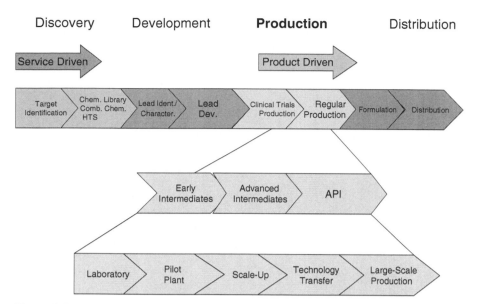

Figure 2.1   Drug development stages.
HTS, high-throughput screening; API, active pharmaceutical ingredient.
Source: Lonza

Note: As both contract research organizations and laboratory chemical suppliers provide primarily services, their revenues are excluded from the total size of the fine chemical business, as discussed in Chapter 9.

2.2   FINE CHEMICAL/CUSTOM MANUFACTURING COMPANIES

Fine chemicals are produced either in-house by pharmaceutical or other specialty chemical companies for their captive needs, or as sales products by fine chemical companies. The latter, that is, the merchant market, accounts for about one-third of the total (captive + merchant) value of $85 billion (see Table 9.2). In business transactions, CM prevails over straight trading of standard products. Whereas the pharma industry constitutes the dominant customer base for most fine chemical companies, some have a significant share of products and services for the agrochemical industry. Examples are Archimica, Saltigo (both in Germany) and Hikal, India.

Only a minority of the companies have been founded with the specific intent of producing fine chemicals, for example, F.I.S. and Flamma, Italy; Hovione, Portugal; Synthetech, USA; and Divi's Laboratories, India. Some have developed by forward integration from fertilizers and chemical commodities, for example, BASF (Germany), Daicel (Japan), Jubilant Organosys (India), and Lonza (Switzerland), or from coal mining, for example, DSM (The

TABLE 2.1   Structure of the Fine Chemical Industry

Type	N	Sales ($ million)	Characteristics
Big	≈10	>250	Divisions of large publicly owned enterprises; >3 sites, total reactor volume >500 m³; short-term profit maximization
			Large in-house capabilities (R&D, manufacturing, marketing)
			Growth by acquisitions
Medium	≈50	100–250	Publicly or privately owned "pure players"; quick, focused decisions; long-term profit optimization. Adequate technology toolboxes, 1–2 sites in the home country, limited global marketing organization
			Growth by reinvesting profits
Small	>500	<100	Focus on niche technologies (azide chemistry, halogenations, phosgenation, peptide synthesis, HPAI)
			Typically privately owned

Netherlands) and UBE (Japan). Others have emerged from diversification, for example, Evonik (Germany) from noble metals; Dottikon Exclusive Synthesis (Switzerland); SNPE (France) from explosives; and Novasep (France) from separation equipment manufacturing. Finally, some originated from pharmaceutical companies, for example, Fermion (Finland), Nicholas Piramal (India), and Siegfried (Switzerland).

Several large pharmaceutical companies market fine chemicals as subsidiary activity to their production for captive use, for example, Abbott, USA; Bayer Schering Pharma, Boehringer-Ingelheim, Germany; Daiichi-Sankyo (after the takeover of Ranbaxy), Japan; Genzyme, USA; Johnson & Johnson, USA; Merck KGaA, Germany; Pfizer (formerly Upjohn), USA.

The fine chemical industry is very fragmented. The top 20 companies have a market share of merely 20%, the top 40 have 30%. In comparison, the top 20 pharma companies have a market share of more than 50%. They also vary substantially in size. The largest ones have sales of more than $500 million; the smallest ones, a few million dollars per year (see Table 2.1). The leading companies are typically divisions of large, diversified chemical companies. The majority are located in Europe, particularly along the axis Amsterdam (The Netherlands)/Basel (Switzerland)/Florence (Italy) and in the United Kingdom.

In terms of size, resources, and complexity of the chemical process technologies mastered, the fine chemical companies can be broadly divided into three categories (see Table 2.1).

The *top tier*, about 20, have sales in excess of $250 million per year (see Table 2.2. Most are divisions or business units (b.u.'s) of large, multinational companies. There are only a few pure players in fine chemicals. All are active

TABLE 2.2 Leading Fine Chemical Companies (Resp. Divisions)

Company		Division		Fine Chemicals	
Name	Sales 2009 ($ million)	Name	Sales 2009 ($ million)	Business Type	Sales 2009 ($ million)
Albemarle	2005			Fine chemicals (a.o., ibuprofen) →	500[a]
Aubinda	665			API-for-generics →	340
BASF	73,000	Care Chemicals	3405	Services	550[a,E]
Boehringer-Ingelheim	18,300	Industrial Chemicals	1100	HMW/LMW=84/16	950
Cambrex	235			Fine chemicals →	240
CSPC, Shijiazhuang Pharmaceutical Group	1500[b]			Pharma Fine Chemicals (APIs), a.o. HIV/ AIDS, sartans →	550[E]
Divi's Laboratories	250			API/CM=70/30 →	250
Dr. Reddy's	1370			Pharmaceuticals and Active Ingredients (PSAI) →	370[a]
DSM	11,300	Pharma	1022	PFCs, CM	850[a,E]
Evonik-Degussa	18,900	Health and Nutrition	2300	Exclusive synthesis	350[E]
F.I.S.	230			API/CM=20/80 →	230
Johnson Matthey	12,500			Fine chemicals →	350
Jubilant Organosys	800E	Pharma and Life Sciences	650	CRAMS	470
Lonza	2600	Custom Manufacturing	1370	HMW/LMW ~55/45	1370
Merck KGaA	11,200	Performance and Life Science Chemicals	1730	Life Science Solutions	580
NCPC, North China Pharmaceutical	718	PFCs (bulk and formulated APIs), antibiotics, vitamins, biopharmaceuticals, etc.			300[E]
Pfizer	50,000	Pfizer Global Manuf.	n/a	Pfizer CentreSource	250[E]
Lanxess	7280	Advanced Interm.	1600	Saltigo	550[E]
Sigma Aldrich	2148	SAFC	601	SAFC/delivery systems →	570[E]
Sumitomo Chemicals	17421			Fine chemicals and polymer additives →	730
Total					**~10,000[a]**

[a] Part of the sales derive from non-fine chemical activities, for example, formulated generics, catalysts, excipients.

[b] 2008.

[E] Estimate (no figures published by the company).

HMW, high molecular weight; LMW, low molecular weight.

11

both in standard products (especially API-for-generics) and CM. They have extensive resources in terms of specialists, plants, process knowledge, backward integration, international presence, and so on. Their manufacturing plants spread over many different locations. Many have grown to their present size through massive acquisitions. Examples are Albemarle, BASF, DSM, Evonik-Degussa, Johnson Matthey, and Pfizer CentreSource.

Very few of the large companies (resp. divisions) listed are pure players in fine chemicals. Considering only fine as defined in Section 1.1, their market share varies between 1% or less for BASF and Pfizer all the way to 100% for F.I.S. and Siegfried. Custom manufacturing typically accounts for at least half of total sales; the balance is standard products, primarily API-for-generics. The European and U.S. pharmaceutical industry is the major customer base. Sales of the majority of the Western companies have been essentially flat in the period 2002–2009. In contrast, strong sales growths were achieved by the Far Eastern companies. Among the Western companies, Boehringer-Ingelheim, Lonza, and SAFC developed above average; Boehringer-Ingelheim and Lonza grew due to their blooming biopharmaceutical businesses, while SAFC started relatively recently from a low level and grew by acquisitions. Because of the high incidence of fixed costs, profits of most firms were even more affected than sales revenues (see Fig. 8.2). Businesses grown by acquisitions during the "irrational exuberance" of the late 1990s suffered most. Thus, Bayer Fine chemicals', Clariant's, and Rhodia Pharma Solutions' sales eroded, and losses were reported prior to the spin-off as, respectively, takeovers by Saltigo, Archimica, and Shasun.

A ranking of the companies according to the size of their fine chemical business is not possible because each of them has a different classification and a different definition of the term "fine chemicals."

Within the world's number 1 chemical company, BASF (sales 2009: $73 billion), fine chemicals are part of the division *care chemicals*. The division had revenues of €3.405 billion/$4.4 billion in 2009 and comprises both *specialties* (e.g., surfactants, UV filters, chelating agents, and excipients) and *fine chemicals*, namely the vitamins A, B_2, C, and E, carotenoids, enzymes, APIs, for example, caffeine, ibuprofen, and pseudoephedrine, and exclusive products for the pharma industry. Financials are only broken down to the divisional level. It is estimated, that fine chemicals account for not more than 15% of the division's sales.

A similar situation is encountered with Albemarle. The $500 million sales of the fine chemicals segment comprise, apart from exclusive synthesis and APIs (ibuprofen, naproxen), *commodities* such as bromine and derivatives, potassium and chlorine chemicals, aluminium oxide, tertiary amines, and *specialties* such as biocides, oilfield chemicals, paper sizers and fillers, matting agents for paints and coatings, so that the factual fine chemicals business accounts for not more than $100–150 million.

The *second tier* consists of several dozens of *midsized* companies. They include both independents and subsidiaries of major companies. A number of these companies are privately owned and have grown mainly by reinvesting the profits. Examples are Bachem, Switzerland; Dishman, India; F.I.S. and Poli Industria Chimica, Italy; Hikal, India; and Hovione, Portugal. The portfolio of the midsized companies comprises both exclusive synthesis and API-for-generics, and sales are in the range of $100–$250 million per year.

Finally, there are hundreds of *small independents* with sales below $100 million per year. Most of them are located in Asia.

Each of the three tiers accounts for approximately the same turnover, namely about $10 billion.

All big and medium-sized fine chemical companies have current Good Manufacturing Practice (cGMP)-compliant plants that are suitable for the production of pharmaceutical fine chemicals (PFCs). With the exception of biopharmaceuticals, which are manufactured by only a few selected fine chemical companies, primarily Boehringer-Ingelheim's biopharmaceuticals division, Lonza, and Nicholas Piramal (formerly Avecia), the technology toolboxes of all these companies are similar. This means that they can carry out practically all types of chemical reactions. They differentiate on the basis of the breadth and quality of the service offering. Most of the medium-sized fine chemical companies are located in Europe, particularly in France, Germany, Italy, the United Kingdom, and Switzerland. Italy and Spain, where international drug patent laws were not recognized until 1978 and 1992, respectively, are strongholds of API-for-generics (see Section 12.2.2). Midsized fine chemical companies have traditionally performed better than large ones. Because of their inherently more attractive offering, this situation will be accentuated in the future.

Large fine chemical companies, in contrast to midsized and small ones, are characterized by

- *Lack of Economy in Size.* As most fine chemicals are produced in quantities of not more than a few 10 tons per year in multipurpose plants, there is little or no economy of size (see Section 5.1). The reactor trains of these plants are similar throughout the industry. Regardless of the size of the companies, their main constituents, the reaction vessels, have a median size of the $4–6\,m^3$. Various products are made throughout a year in campaigns. Therefore, the unit cost per cubic meter per hour practically does not vary with the size of the company.

  Contrary to what one would expect, size also frequently does not result in a reduction of risks. More likely it leads to an increasing dependence on very few key customers and products.

- *Dichotomy between Ownership and Management.* The company's shares are listed on stock exchanges, and their performance is scrutinized by the financial community, which has a short-term view. Postponement of a single important shipment can affect a quarterly result. In the small and midsized companies, the owners typically are the major shareholders, often members of the same family. Their shares are not traded publicly and fluctuations in their financial performance are more easily coped with.

- *Complicated Business Processes.* Flexibility and responsiveness are in jeopardy. Customer complaints, for instance, are difficult to handle in a straightforward manner: Before the big company can determine in which plant the defective batch had been produced, the small company would have settled the complaint.

A business development manager, who had moved from a big to a small fine chemical company, stated that "At my previous employer, it took me three trips to the USA and one to Italy just to determine, which step should be produced at which site."

- *Heterogeneous Portfolio of Small Companies, Accumulated over Time through M&A Activities.* The key functions, such as production, R&D, and M&S, are located on different sites, often in different countries.
- *Cohabitation with Other Units.* Running side by side units of different size, serving different markets, and using different technologies is a daunting management task.

Customers prefer to do business with *midsized* companies because communications are easier (they typically deal directly with the decision maker)— and they can better leverage their purchasing power. They also do not want to depend too much on single, large, and powerful suppliers (where they do not know "who is in charge").

The *small* fine chemical companies have only limited capabilities and often specialize in niche technologies, such as reactions with hazardous gases (e.g., ammonia/amines, diazomethane, ethylene oxide, halogens, hydrogen cyanide, hydrogen sulfide, mercaptans, ozone, nitrous oxides, phosgene). Their small size, however, is not necessarily a disadvantage. The minimum economical size of a fine chemical company depends on the availability of infrastructure. If a company is located in an industrial park, where analytical services, utilities, safety, health, and environmental (SHE) services, and warehousing are readily available, there is practically no lower limit. New fine chemical plants have come onstream mostly in Far East countries over the past few years (as of 2006), but their annual turnover rate rarely exceeds $25 million.

Midsized and small fine chemical companies were also impacted by the slump in demand for new PFCs. By and large, they fared better than did the large companies, the drawbacks of which have been described above. Selected small and midsized fine chemical companies are listed here:

- *Europe:* Belgium (Omnichem [Ajinomoto]), Czech Republic (Interpharma, Synthesia), Denmark (Axellia Pharmaceuticals, Polypeptide), Germany (Archimica, Chemische Fabrik Berg, Corden Pharma, Girindus, Organica Feinchemie Wolfen, Pharma Waldhof, Wacker Chemie), Finland (Fermion,* KemFine), France (Axyntis, La Mesta Chimie Fine [AETGroup], Isochem [acquired by Aurelius from SNPE in 2010] Minakem, Orgasynth, PCAS, Simafex, Synchem], Hungary (Egis Pharma,* Gedeon Richter*), Israel (Chemada, Chemagis), Italy (Dipharma, Erregierre, F.I.S. (see photo 1 in the insert), Flamma, Recordati,* Zach Systems), Latvia (Olainfarm), Norway (Borregaard), Poland (Polpharma), Portugal (Hovione), Spain (Esteve Quimica,* Medichem, Uquifarma [Yule Catto]), Sweden (Dupont Chemoswed), Switzerland (Bachem, CU Chemie Uetikon, Cilag [J&J], Dottikon ES, Helsinn,* Siegfried]), Turkey (Atabay), and United Kingdom (Aesica Pharmaceuticals,* Contract Chemicals, Excelsyn, Robinson Brothers).

- *North America:* USA (Albany Molecular Research, Albemarle, AMPAC Fine Chemicals, Ash Stevens, Cambridge Major Laboratories, Codexis, Ferro Pfanstiehl Pharmaceuticals, SAFC, Synthetech [acaquired by WR Grace in 2010]), Canada (Delmar).

- *Far East:* China (Zhejiang Huayi Pharmaceutical, Zhejiang Hisun Pharmaceutical), India (Cipla,* Divi's Laboratories, Dishman, Hikal, Jubilant Organosys, Nicholas-Piramal, Ranbaxy,* Suven), Japan (API Corp., Daicel F.C., Kuraray, Nippoh, Nippon Gohsei, Takasago, UBE F.C.), South Korea (Hanmi Pharm,* Samchully, SK Energy & Chemicals), Taiwan (Scinopharm, Syn-Tech).

Note: *Fine chemical companies that are also active in formulated pharmaceuticals.

A comprehensive list of about 1400 fine chemical companies (including traders) can be found in the "event catalogue" of the CPhI exhibition (see Appendix A.1). A category of mostly European and American small fine chemical companies do not have manufacturing plants and concentrate on research and process development (see Chapter 3).

2.3 CONTRACT RESEARCH ORGANIZATIONS

CROs provide services to the life science, especially pharmaceutical, industries along product development. There are more than 2000 CROs operating worldwide, representing revenues of more than $20 billion. One distinguishes between *"product" CROs* and *"patient" CROs*.

Patient CROs, aka *chemical CROs* are providing primarily process research and development services. Their tasks are described in Table 2.3. Whereas the production sites of contract manufacturing organizations (CMOs) are multi-purpose plants, allowing for the production of tens to hundreds of tons of fine chemicals, the "workplaces" of patient CROs are the test persons (volunteers) for the clinical trials and those of the product CROs are the laboratory benches. An overlap between the latter and CMOs exists with regard to pilot plants (100 kg quantities), which are part of the arsenal of both types of enterprise. Companies offering CRAMS, aka one-stop-shops, also exist. They are described at the end of this chapter.

The offerings of *patient CROs*, aka *clinical CROs*, comprise more than 30 tasks addressing the clinical part of pharmaceutical development at the interface between drugs, physicians, hospitals, and patients. Examples are the clinical

TABLE 2.3   Tasks of "Product" Contract Research Organizations

Task	Description
Sample Preparation	
Synthetic PFCs	Laboratory preparation of PFCs, impurities, metabolites, etc.
Natural products	Product extraction, purification, and characterization
Process Development	
General	Upgrading of laboratory procedures to economically and ecologically viable industrial-scale manufacturing processes[a] (including examination of process parameters)
Route screening	Evaluation of the most suitable synthetic or biotechnological route (mostly by literature search)
Proof of principle	Confirmation of selected route based upon economic and quality criteria, equipment specifications, etc.
Sample preparation	Reference and impurity standards of PFCs
Safety and toxicology studies	Hazard and toxicological (including genotoxicity) tests required for industrial-scale manufacture
Analytics	Analytical method development and validation
Process research	• Process optimization
	• Definition of the parameters for industrial-scale manufacture
	• Method validation
	• Stability studies
Regulatory affairs	• Production permits
	• API submissions (IND, NDA support)
Scale-up (kilogram-laboratory/ pilot and industrial scale plant production)	• Confirmatory testing of the process
	• Preclinical and clinical trial quantities
	• Validation manufacturing (Phase III and beyond)

[a] See also Chapter 6.
IND, investigational new drug; NDA, new drug approval (or applications).
Source: Jan Oudenes, Alphora Research, private communication.

development and selection of lead new drug compounds, planning, monitoring, and analyzing Phase I–IV clinical studies, ADMET (absorption, distribution, metabolism, excretion, and toxicity) studies, development of diagnostic kits, and devising and executing complex marketing programs for launching new drugs. The patient CROs have rudimental capabilities for synthesizing PFCs at best. As clinical trials, including the associated data management, represent the largest expense in pharmaceutical research, the market for patient CROs is larger than for their product counterparts. Thus, the sales of the top-tier firms

TABLE 2.4   Major "Product" and "Patient" Contract Research Organizations

	"Product" CROs
North America	Alphora, Canada (85[b]); AMRI, USA (229 million[a]/1200[b]); Aptuit, USA ($2.7 billion[a]/2700[b]); Cambridge Major, USA (175[b], 50[c]); ChemBridge (300[b]); Delmar, Canada (70[b]); Innocentive, USA (N/A): Irix Pharmaceuticals, USA (150[b]/100[c]); NAEJA, Canada ($150 million[a]/120[b]); PharmEco, USA (120[b])
Europe	Carbogen-Amcis, Switzerland (450[b]/120[c]); Chemcomm, Germany (50[c]); ChemDiv, Russia (N/A); Clauson-Kaas, Denmark (30[b]); CRS Clinical Research Services, Germany (150[b]) Enamine, Russia (300[b]); Excelsyn[d], UK ($15 million[a]/60[b]); Girindus, Germany (75[b]); Nerviano Medical Sciences, Italy (700[b]); Onyx- Scientific, UK (40[b]/36[c]); Recipharm, Sweden ($265 million[a]/1900[b]); Serichim, Italy (24[b]/14[c]); Solvias, Switzerland ($60 million[a], 280[b])
Asia	Acoris, India (50[b]); Aptuit Laurus, India (700[b]); Biocon/Syngene, India (3000[b]); BioDuro, China (675[c]); Chembiotek, India (300[c]); Chempartner, China (2000[b]/1250[c]); Medicilon, China 400[b]/240 [c]); NARD Institute, Japan (N/A); Pharmaron, China (700[b]); ProCitius, India ($350 million[a]); Riken, Japan ($23 million[d]); WuXi AppTec, China ($270 million[a]/>4000[b], 2800[c])
	"Patient" CROs
North America	Charles River Laboratories, USA ($1.2 billion[a]/8100[b]); Covance, USA ($1.9 billion[a]/9800[b]); Icon, Ireland ($0.9 billion[a]/7200[b]); Parexel, USA ($1 billion[a]/9000[b]); PPD, USA ($3 billion[a]/10,500[b]); Quintiles Transnational, USA ($2+ billion[a]/23,000[b])
EU	Biotrial, France (250[b]); Focus, Germany (120[b]); Pharmalog, Germany (50[c]); TFS Trial Form Support, Sweden ($50 million[a]/430[b])
Asia	Chemizon, South Korea (150[b]); gvk BioSciences, India (1400[b]); TCG Lifesciences, India (1000[b]); Virginia CRO, Taiwan (N/A)

[a] Sales (2009).
[b] Total staff.
[c] Scientists.
[d] Life Science unit only, Excelsyn was acquired by AMRI for $19 million in February 2010.

are in the $1–$2 billion range, whereas the largest product CROs have revenues of a few $100 million (see Table 2.4). The distinction between the two types of CROs has become blurred recently. Thus, Aptuit, USA, is very diversified. Apart from clinical research services, its activities also comprise clinical operations, small-scale manufacturing of high potency and cytotoxic APIs, drug formulation, packaging, and distribution.

Major product and patient contract research organizations are listed in Table 2.4.

There are about 50–100 "product" CROs in developed countries, either stand-alone companies or divisions of larger chemical companies, with a widely differing degree of width and depth of their offering. The typical history of a

CRO begins with a chemist working on a thesis and trying to make some pocket money by preparing samples for a life science company. Gradually, the chemist's part-time job develops to a full-time activity. Colleagues are employed, and a CRO company is founded. Most "product" CROs are privately held and have revenues of $10–$20 million per year or less, adding up to a total business in the range of $1.5–$2 billion.

Major customers for CRO services are the large global pharmaceutical companies. Half a dozen "big pharma's" (Pfizer, GlaxoSmithKline, Sanofi-Aventis, AstraZeneca, Johnson & Johnson, and Merck) alone absorb about one-third of all CRO spending. As for CMOs and also for CROs, biotech start-up companies with their dichotomy between ambitious drug development programs and limited resources are the second most promising prospects after "big pharma" (see Section 12.3).

> An example of a leading chemical CRO is Albany Molecular (AMRI). It had total revenues of $190 million in 2008 (CRO, 34%; CMO, 51%, royalties, 15%). AMRI does organic synthesis and chemistry development, supported by computational chemistry for molecular modeling, with computer-assisted drug design. Furthermore, it offers different types of libraries: custom, semi-exclusive, focused, and natural products. Finally, AMRI conducts its own proprietary R&D aimed at licensing preclinical and clinical compounds.

Asian, especially Chinese and Indian, companies are emerging as low-cost contract research providers. In India alone, there are more than 20 chemical CROs. The largest is Syngene, a division of Biocon, with over 3600 employees and sales of $360 million in 2008/2009, followed by gvkBio, Chembiotek, and ProCitius (Sanmar). Under the name of Acoris, Hikal runs a research center for 200 scientists in Pune, India (see photos 8 and 18 in the insert). The largest Chinese chemical CRO is WuXi AppTech, Shanghai WaiGaoQiao Free Trade Zone, which was set up in the year 2001 and led by 50 returnees, with 2009 sales of $270 million of R&D services (preparing libraries, carrying out preclinical tests, toxicological studies, process synthesis) and also approximately $30 million of CM. They employ 2700 workers (of which more than 1300 scientists) and are growing 30% per year.

The business of CROs is usually done through a "pay for service" arrangement. Contrary to manufacturing companies, invoicing of CROs is not based on unit product price, but on full-time equivalents (FTEs), that is, the cost of a scientist working 1 year on a given customer assignment. For further details, see Sections 12.5 and 12.6.

Key reasons for outsourcing R&D activities are to

- Allow pharma companies to develop drugs faster to maximize patent protection and secure marketplace advantage.
- Contain cost—since only one in three drugs recovers its cost of development, the pharmaceutical industry needs to find ways to increase the

supply of drug candidates and at the same time reduce development costs. Outsourcing to top-tier CROs can accomplish both of these objectives.

• Deferring internal increases in headcount and expenses.
• Buffering demand peaks for in-house R&D services.

Contract Research and Manufacturing Organisations (CRAMs) are hybrids combining the activities of CROs and CMOs. Their history is either a forward integration of a CRO, which adds industrial scale capabilities (an early example is Suven, India, a recent one Cambridge Major in the United States), or backward integration of a CMO. Examples of CRAMs are DSM, The Netherlands; Lonza, Switzerland; Nicholas Piramal, and Jubilant Organosys, India. It is questionable, though, whether one-stop shops really fulfill a need. The pros and cons are summarized in Table 2.5.

The first "pro" entry in Table 2.5, "Chance to establish a relationship with a drug company early on," is particularly debatable. Most new drugs fail in early-stage development. The situation has worsened over the years. Nowadays, even for developmental drugs in Phase II, the probability of reaching the market is less than 10%. Furthermore, as there is little repeat business, and as in big pharma different functions are in charge of sourcing laboratory chemicals as opposed to outsourcing chemical manufacturing, sample orders only rarely evolve to industrial-scale supplies.

An example in point is Johnson Matthey, the world's largest supplier of opiates. The products are obtained by plant extraction, which is the company's core competence. JM acquired the "chemical" CRO PharmEco with the intent to offer a one-stop shop capability. As PharmEco was primarily involved in synthetic chemistry, it is difficult to come across a synergy between the small- and the large-scale business.

TABLE 2.5   Pros and Cons of the "One-Stop Shop" Concept

Pros	Cons
Fine Chemical/Custom Manufacturing Company	
• Chance to establish a relationship with a drug company early on • Higher overall added value	• In >90% of cases, projects are stopped at the lab sample stage • Need to master two different skills: "quick and dirty" lab scale versus economically viable and ecologically safe large-scale production
Pharmaceutical Company	
• Reduction of number of suppliers	• In contrast to the policy of selecting specialists for each step of drug development • Overdependence on one supplier

Actually, the large fine chemical companies consider the preparation of samples more as a marketing tool (and expense ...) rather than a profit contributor. In order to avoid some of the pitfalls, it is advisable to manage the CRO business as a separate unit. Also, the location should preferably be separate to ensure that there is greater accountability and ability for it to operate as a stand-alone business. It should be determined by availability of talent, proximity of universities, and accessibility. Also, the implementation of a rigorous confidentiality and intellectual property (IP) safeguard plan is mandatory to protect both the customer and the CRO. Hovione (Portugal) is an example in case. Its Technology Transfer Centre, which is in charge of CRO, is located in New Jersey.

> Innocentive, Andover, MA, USA, is a particular, virtual CRO. It offers companies the possibility to post research problems, such as a synthesis for a new compound anonymously on the Internet. Its website now connects more than 95,000 registered scientists around the world. Financial incentives up to $100,000 are paid to successful problem solvers. The success rate runs at about 35%.

2.4 LABORATORY CHEMICAL SUPPLIERS

Before the life science industry, colleges and universities, medical research institutions, hospital research labs, government agencies, and other facilities can initiate any chemical research activity, they need chemicals, solvents, and laboratory equipment. The laboratory chemical suppliers provide these items. Their combined revenues are about $10 billion. The key success factors are speed, ease of ordering, and number and quality of the products. A laboratory chemist, or team leader, must be in a position to order samples online, and to receive them quickly and in the right quality. Regarding the size of the offering, the five top-tier companies are

- *Life Technologies*, USA (*formerly* Invitrogen Corporation *and* Applied Biosciences). Sales were >$3 billion in 2009 and the catalog lists 50,000 products. "Our systems, consumables and services enable researchers to accelerate scientific exploration."
- *Sigma Aldrich*, USA. The company had total sales of $2.1 billion in 2009, of which laboratory chemical supplies account for about 70%. Apart from offering 100,000 small and big molecules, including cell culture media, it has also 30,000 items of laboratory equipment available. Under the name "SAFC Pharma," a PFC (including high-potency active ingredient [HPAI]) CM business is operated. It represents 30%, or $630 million of total sales.
- *Thermo Fisher Scientific*, USA (*formerly* Fisher Scientific *and* Thermo Electron). The number one company in this field (sales $9 billion [2009^E)]) supplies biochemicals and bioreagents; organic and inorganic chemicals (of which 15,000 are fine organic chemicals); sera; cell culture media;

sterile liquid-handling systems; microbiology media and related products; scientific consumable products, instruments, and equipment.

- *Tokyo Kasei Kogyo Co., Ltd.*, Japan. This company offers 20,000 products.
- *VWR International* (owned by CDRV Investors, USA). The number 2 company had sales of $3.8 billion (2009^E). The distributor of laboratory supplies represents 5000 manufacturers, of which Merck KGaA's "analytics and reagents" and VWR's "scientific products" are the most prominent ones. According to company information, VWR offers 750,000 products, including small- and big-molecule laboratory chemicals.

Online ordering is possible from all these companies. With a market share of about 45%, Sigma Aldrich is the market leader in e-commerce. Apart from the top five, there are many laboratory chemical suppliers with smaller catalogues geared at specific needs, such as BioCatalytics, which offers a ketoreductase kit with about 100 enzymes, or Chiral Technologies, a division of Daicel, Japan, which offers a range of 175 immobilized and coated polysaccharide chiral stationary phases for use with high-pressure liquid chromatography (HPLC), supercritical fluid chromatography (SFC), and simulated moving bed (chromatography) (SMB) equipment. A selection of *N*-heterocyclic compounds, especially azaindoles, naphthyridines, pyridines, and pyrrolidines, is offered by Adesis, USA. Peptide building blocks are offered by Bachem, Switzerland (9000 products); Polypeptide, Denmark (150 products); Senn Chemicals, Switzerland (1500 products); and Synthetech, USA (250 products); and CBL-Patras, Greece offers 400 different types of "Barlos resins" for solid-state peptide synthesis.

2.5 MERGERS AND ACQUISITIONS (M&A)

The M&A scenario for the fine chemical industry has changed completely over the past 10 years. Around the turn of the millennium, large specialty chemical companies, plagued by an escalating commoditization of their product portfolios (and a collateral profit erosion), such as Akzo Nobel, Avecia (formerly I.C.I.), Clariant, Dow, DSM, Eastman Chemicals, Evonik-Degussa, Honeywell, Rhodia, Shell Chemicals, and Solvay endeavored improving their overall performance by acquiring fine chemical companies. Investment bankers had promoted them as offering a better profit and growth potential than their traditional activities. During this period of "irrational exuberance," high earning multiples were paid for acquisitions. The climax was reached when Clariant paid a price/earnings (P/E) ratio of 28 for BTP, UK, had to write down the acquisition price, $1.7 billion in 2000, almost totally in subsequent years, and finally sold it to International Chemical Investors Group (ICIG) (see below).

By 2005, the situation inverted from a seller's to a buyer's market. As most of their acquisitions had not met with the expectations, the specialty chemical companies began to divest their fine chemical units. At the same time, big pharma companies started offering many of their chemical manufacturing plants for sale as part of their restructuring programs (see Section 16.1). As a large number of plants came on the block, P/E ratios nose-dived from more than 20 to less than 10. Three categories of M&A deals now prevail: (1) *Western financial investors* have been playing an increasingly important role in the acquisition scene in the recent past (see Table 2.6). Their positive valuation is based on the attractiveness of the buzz word "life sciences" in general and on big pharma's paradigm change from opportunistic to strategic outsourcing, which should provide growth opportunities for the fine chemical/CM industry worldwide. ICIG, Frankfurt/Main is the leading player. Since its inception in 2004, ICIG has acquired altogether 15 fine chemical units both from pharma and fine chemical companies with a combined turnover of approximately $1 billion. The portfolio comprises a.o. Albemarle's Thann/Mulhouse works, Astra Zeneca's Plankstadt/Schwetzingen plant, renamed Corden Chemicals, two Cambrex facilities in Belgium and the United Kingdom, respectively, Clariant's CM business, several businesses of Rütgers Werke, including WeylChem, Solvay's Synkem business (originally part of Laboratoires Fournier), and Synthacon, Frankfurt. (2) *Far East*, mainly Indian, chemical, and life science companies also have become active (see Appendix A.10). Apart from the attractive prices, their motivators are an upgrading of their value-added chain, acquisition of know-how, and creation of a foothold in the backyards of Western life science companies for facilitating business (see also Section 14.2). Apart from these rational justifications, a kind of herd or pack instinct is also gaining ground ("If xy does it, it must also be good for me"), similar to the "irrational exuberance" of the Western M&A activities around the turn of the century. The opposite, namely acquisition of Asian companies by Western life science (fine chemical, pharma, and agro) companies, would also make sense, as it would allow an optimization of the cost structure. However, both the high market capitalization and the high stakes held by founder shareholders prevent any hostile bid in most cases. Actually, the acquisition of Matrix (see also Section 10.1) by Mylan, one of the top 10 U.S. generic companies, is the only major foreign takeover of an Indian fine chemical company. Mylan's rationale was the backward integration of the supply chain by a strong API-for-generics manufacturer. Mylan paid $736 million for 71.5% of Matrix's shares, corresponding to a striking P/E ratio of 22.

Category (3) are *fine chemical companies looking out for plants offered by big pharma* as part of their manufacturing restructuring programs (see Section 16.1). It enables the company to dispose of assets without being forced to lay off employees. When pondering acquisition opportunities, they first have to make a choice between the offerings from big pharma, specialty chemical firms, and stand-alone plants. As a matter of fact, there are substantial differences: Contrary to the quality standards of stand-alone fine chemical plants, which typically are "bare operational minimum," plants built by pharma companies are often "gold-plated." Fine chemical plants, which are part of specialty chemical firms, take an intermediate position. As big pharma offer a huge

TABLE 2.6 Acquisitions of Fine Chemical Companies or Divisions by Private Equity Firms

Company	Activity	Investor	Year	Financials ($ million)	
				Sales	Acq. Price
VWR Int'l. (Merck KGaA)	Laboratory chemical supplies	CDRV Investors (Madison Dearborn)	2002	3700	N/A
KemFine (Kemira Oy)	Fine chemicals (mainly agro)	3i	2004	120E	N/A
Archimica (Clariant)	Fine chemicals (mainly APIs)	TowerBrook Capital Partners	2006	170	N/A
CABB[a]	Sulfuryl and thionyl chlorides, CM	AXA Private Equity	2006	450	N/A
Cambrex HH[b]	Fine chemicals (mainly APIs)	International Chemical Investors Group (ICIG)	2006	42	N/A
Clariant PFC	Agro and pharma Fine chemicals	TowerBrook Capital Partners	2006	175	90
Vertellus[c] (Reilly Tar, Rutherford)	Pyridine and picolines, DEET, castor oil der.	Arsenal Capital Partners	2006	400	314[c]
Cambridge Major Labs	Contract res. and manufacturing	Arlington Capital Partners	2007	N/A	N/A
Novasep (Rockwood)	FCs (a..o.N_3 chemistry), SMB	Gilde Buy-Out Partners Banexi Cap. Part.	2007	350	530
Axellia (Alpharm)	Injectable antibiotic APIs	3i	2008	N/A	395
Merck & Co. Cherokee Plant	Fine chemicals (mainlyAPIs)	PRWT Services	2008	N/A	N/A
WeylChem (Rütgers)	Fine chemicals (>50% agro)	International Chemical Investors Group (ICIG)	2008	270E	N/A
Alzchem (SKW Troisdorf)	NCN chemistry	bluO Luxemburg	2009	400E	N/A
Rohner-Chem (Arques)	Fine chemicals	bluO Luxemburg	2009	N/A	N/A
Miteni (Mitsubishi)	Fluoro-organics	International Chemical Investors Group (ICIG)	2009	50E	N/A

TABLE 2.6   *(Continued)*

Company	Activity	Investor	Year	Financials ($ million)	
				Sales	Acq. Price
Keata Pharma Pfizer Arnprior	Fine chemicals (mainly APIs)	Pillar5, Canada			
Isochem (SNPE)	Phosgene derivatives subst. amino acids	Aurelius	2010	160	11
Polichimica	Fine chemicals (mainly APIs)	Mandarin Capital Partners (Italo-Chinese)	2010	50	N/A

[a] Clariant Acetyl Building Blocks; The portfolio comprises the former Clariant plant in Sulzbach, Germany, the former Säurefabrik Schweizerhalle plant in Switzerland, and the former Karnavati Rasayan in India.

[b] ICIG acquired the Cambrex Human Health plants in Cork, Ireland and Landen, Belgium, and made them part of its new company, Corden PharmaChem.

[c] Arsenal acquired. Rutherford Chemicals for $64 m from Cambrex in 2003; Reilly Tar for $250 m from the Reilly family in 2005.

Note: Names in parentheses (…) designate previous owners.

discount for the superb plants they want to eliminate, the high initial construction cost per se is not a financial problem. However, as the plants were run as cost rather than as profit centers, it is very difficult to implement a culture of lean production. As soon as the deal is made, the new owners must turn previously in-house, often single-product plants into customer-focused multipurpose operations. Lonza, for example, tried it for years and finally decided to close both its 150-m³ fine chemical plants in Conshohoken, PA and in Los Angeles. Not surprisingly, in contrast with the large offer (Pfizer alone already has eliminated 57 plants. On top, also part of the 43 plants resulting from the Wyeth acquisition will be divested, see Section 16.1), few acquisitions of big pharma manufacturing sites by fine chemical companies have been made so far (see Table 2.7).

The exception to the rule is Aesica founded 2004 by a management buy-out of BASF's Cramlington, UK plant. It has expanded rapidly by acquiring plants divested by big pharma (a.o. Merck Inc.'s Ponders End plant and Abbott's Queensborough formulation plant).

Further important considerations for the acquiring party are: What kind of supply contracts are offered to ascertain at least a base load? The sale is often sweetened by a "sell and lease-back" contract. It assures continuing product supplies to the seller and provides a base load business for the acquiring party.

TABLE 2.7   Pharmaceutical Companies' API Plants Acquired by Fine Chemical Companies

Acquirer	Plant Acquired	Activity	Year	Remarks
Inyx,[a] USA	Sanofi-Aventis' Puerto Rico	Dosage forms	2005	Acquisition price: $20 million
Aesica Pharmaceuticals, UK	Merck's Ponders End, London	PFC	2006	Prod. cap. 300 mtpa 75 employees
	Abbott's Queensborough, UK	PFC, formulation	2007	450 employees
Abraxis Bioscience, USA	Pfizer's Cruce Davila, PR	APIs, a.o. Celebrex	2007	Acq. price $32 million; 450 employees
Minakem, France	AstraZeneca's Dunkirk plant	APIs, a.o. omeprazole	2009	Reactor vol. 250 m³
Corden Pharma, Germany	GSK's Annan, Scotland plant	PFCs,	2009	
Hovione, Portugal	Pfizer's Loughbeg, Cork, Ireland	PFC, spray drying	2009	Reactor vol. 430 m³
Evonik-Degussa, Germany	Eli Lilly's Tippecanoe Laboratories, USA	PFC, fermentation	2009	Reactor vol. 800 m³ 650 employees

[a] Insolvent.

Even if product offtakes decrease over the contract period, they grant the acquirer a grace period for the introduction of its own products. How generous are the commitments that the new owner has to honor, for example, the labor contracts? Is it advisable to buy dedicated plants and convert them to multi-purpose plants? Such a "flexibilization" is less of a hardware (i.e., engineering) than a software problem: The whole supply chain from project management, raw material procurement, production scheduling, customer relationships, controlling, permits, regulatory compliance, and so on are entirely different.

CHAPTER 3

Products

In terms of molecular structure of fine chemicals, one distinguishes first between low-molecular-weight (LMW) and high-molecular-weight (HMW) products. Molecular weights of LMW life science products typically vary between 200 and 500; those of HMW between 20,000 and 50,000. The LMW fine chemicals, also designated as small molecules, are produced by traditional chemical synthesis, by microorganisms (fermentation or biotransformation), or by extraction from plants and animals. In the production of modern life science products, total synthesis from petrochemicals prevails. The HMW products, respectively large molecules, are obtained mainly by biotechnology processes. Within LMWs, the *N*-heterocyclic compounds are the most important category; within HMWs, it is the peptides and proteins. Significant representatives of the two categories are described in more detail in the following chapters.

3.1 SMALL MOLECULES

As aromatic compounds have been exhausted to a large extent as building blocks for life science products, *N*-heterocyclic structures prevail nowadays. They are found in many natural products, such as chlorophyll, hemoglobin, and the vitamins biotin (H), folic acid, niacin (PP), pyridoxine HCl (B$_6$), riboflavin (B$_2$), and thiamine (B$_1$). In life sciences, 4 of the top 5 proprietary drugs and 5 of the top 10 agrochemicals contain *N*-heterocyclic moieties (see Tables 11.4 and 11.7). Even modern pigments, such as diphenylpyrazolopyrazoles, quinacridones, and engineering plastics, such as polybenzimidazoles, polyimides, and triazine resins, exhibit an *N*-heterocyclic structure.

In the four-membered rings, the β-lactam moiety is part of the classical penicillin and cephalosporin antibiotics. The most prominent example of a drug with a five-membered ring with one nitrogen atom is Lipitor (see Table 1.1). In the five-membered rings with 2N atoms, imidazoles are found both in modern agrochemicals, especially the imidazolinones (e.g., Imazapyr), and

Fine Chemicals: The Industry and the Business, Second Edition, by Peter Pollak
Copyright © 2011 John Wiley & Sons, Inc.

pharmaceuticals, such as antimycotics (e.g., isoconazole, ketoconazole, and miconazole), anticancers (e.g., temozolomide), and antiulcerants (cimetidine and omeprazole). Five-membered rings with 3 N atoms, triazoles or triazolones, are found in other antimycotics (e.g., fluconazole and itraconazole), antivirals (e.g., ribavirin), and antidepressants (e.g., nefazodone hydrochloride). Five-membered rings with four nitrogen atoms, tetrazoles and tetrazolines, are found in a variety of modern antihypertensives ("sartans"; e.g., candesartan, irbesartan, losartan, valsartan), antibiotics (cefotetan and cefazolin), antiallergics (pemirolast and pranlukast), and analgesics (e.g., alfentanil). Pyridine derivatives, six-membered rings with 1 N atom, are found in both well-known Diquat & Chlorpyrifos herbicides, and in modern nicotinoid insecticides, such as imidacloprid. A vast array of pharmaceuticals and agrochemicals are built around a pyrimidine (2 N atoms in 1,3 position) ring structure. Important classes are modern antiviral compounds such as zidovudine and nucleotides (see discussion below). The sulfonamide antibiotics (e.g., sulfadimethoxime and sulfamethazine) set a milestone in modern medicinal chemistry, and—half a century later—the sulfonyl ureas (such as amidosulfuron and bensulfuron-methyl and in modern pest control.

Benzodiazepine derivates, seven-membered rings with 2 N atoms in 1,4 position, are the pivotal structures of the benzodiazepine class of breakthrough CNS (central nervous system) drugs such as Librium and Valium. A widely described benzodiazepine is Seresta (oxazepam), 7-chloro-1,3-dihydro -3-hydroxy-5-phenyl-2H-1,4-benzodiazepin-2-one:

Purines (purine, 7H-imidazo [4,5-d]pyrimidine, $C_5H_4N_4$) and pteridines (pteridine, pyrazino [2,3-d] pyrimidine, $C_6H_4N_4$) are compounds consisting of two fused N-heterocyclic rings. Adenine and guanine are important purines. They are used, for example, as building blocks for nucleotides (see discussion below). Folic acid (a vitamin), methopterin, and methotrexate are typical pteridines.

In bioactive fine chemicals, for example, active substances of pharmaceuticals or agrochemicals, the crystal form of the product also needs to be considered. In order to improve the solubility, salts of active substances have been used early on. More recently, determining and controlling the solid form of drugs and pesticides is a growing area of science and business. If the bioavailability is strongly influenced by the solubility and the dissolution profile, this can have significant consequences and determine whether the compound is further developed. One distinguishes amorphous, crystalline (a given product can exist in different crystal forms), polymorph forms, and solvates. The latest development are co-crystals. They consist of two or more components that are solid at room temperature. They exhibit different physicochemical properties and create a.o. new possibilities for patent protection, respectively extension. A well-documented case is the antifungal itraconazole, in which different carboxylic acid co-crystals exhibit a higher solubility and a faster dissolution rate.

3.2 BIG MOLECULES

Big molecules, also called HMW molecules are mostly oligomers or polymers of small molecules or chains of amino acids. Thus, within pharma sciences, peptides, proteins, and oligonucleotides constitute the major categories.

Peptides and proteins are oligomers or polycondensates of amino acids linked together by a carboxamide group. The threshold between the two is as at about 50 amino acids. Because of their unique biological functions, a significant and growing part of new drug discovery and development is focused on this class of biomolecules. Their biological functions are determined by the exact arrangement or sequence of different amino acids in their makeup. There are 20 naturally occurring amino acids, 8 of which are defined as essential amino acids, namely L-isoleucine, L-leucine, L-lysine, L-methionine, L-phenylalanine, L-valine, L-threonine, and L-tryptophan.

For the synthesis of *peptides*, four categories of fine chemicals, commonly referred to as peptide building blocks (PBBs), are key, namely, amino acids (starting materials), protected amino acids, peptide fragments, and peptides themselves (see also Section 4.1). Along the way, the molecular weights increase from about 10^2 up to 10^4 and the unit prices from about 10^0 up to 10^5 per kilogram. However, only a small part of the total amino acid production is used for peptide synthesis. In fact, L-glutamic acid, D, L-methionine, L-aspartic acid, and L-phenylalanine are used in large quantities as food and feed additives (see Section 11.4). The first synthetic peptide was oxytocin (MW 1007) discovered in 1953 by du Vigneaud, who received the Nobel Prize for this achievement. It was followed by Vasopressin and ACTH (Novartis' Synacthen, see Table 3.1). Nowadays, about 50 peptide drugs are commercialized. A selection of 16 of them is shown in Table 3.1. A shown in column 3, the number of amino acids that make up a specific peptide varies widely. At the low end are the dipeptides. The most important drugs with a dipeptide (L-alanyl-L-proline) moiety are the "-pril" cardiovascular drugs, such as enalapril (shown in the table), captopril, imidapril, and lysinopril. Also, the artificial sweetener Aspartame (N-L-α-Aspartyl-L-phenylanaline 1-methyl ester) is a dipeptide. At the high end there is the anticoagulant hirudin, MW ≈ 7000, which is composed of 65 amino acids.

Apart from pharmaceuticals, peptides are also used for diagnostics and vaccines. The total production volume (excluding Aspartame) of chemically

TABLE 3.1 Typical Peptide Drugs

Peptide	Therapeutic Class	Size[a]	Manufacturing Method
Enalapril	Angiotensin-converting enzyme	2	Chemical synthesis
Thymopentin	Immunoregulator	5	Chemical synthesis
Octreototide	Gastric anti-secretory agent	8	Chemical synthesis
Oxytocin	Oxytocic (labor stimulating)	9	Chemical synthesis
Goserelin	Antineoplastic (hormonal)	9	Chemical synthesis
Leuprolide	Antineoplastic (hormonal)	9	Chemical synthesis
Vasopressin	Antidiuretic and vasopressor hormone	9	Various
Goserelin	Antineoplastic (prostate cancer)	10	Chemical synthesis
Cyclosporin A	Immunosuppressant	11	Extraction from fungus
Somatostatin	Growth stimulant	14	Chemical synthesis
ACTH	Adrenocorticotropic hormone	24	Chemical synthesis
Glucagon, HGF	Anti-hypoglycemic	29	Extraction, recombinant, chemical synthesis
Calcitonin	Calcium regulator (osteoporosis)	32	Various
Fuzeon	Anti-AIDS (fusion inhibitor)	36	Chemical synthesis
Insulin	Antidiabetic	51	Various
Hirudin	Anti-thrombotic	65	Extraction from leeches, recombinant

[a] Number of amino acids.

synthesized, pure peptides is about 1500 kg, and sales approach $500 million on the active pharmaceutical ingredient (API) level and $10 billion on the finished drug level, respectively. The numbers would be much higher, about 10% of total pharma sales, if also peptidomimetics and APIs, which contain peptide sequences as part of a molecule would be included, such as the above-mentioned "... prils" or the first-generation anti-AIDS drugs, the "... navirs." The bulk of the production of peptide drugs is outsourced to a few specialized contract manufacturers, such as Bachem, Switzerland; Chengu GT Biochem, China; Chinese Peptide Company, China; Lonza, Switzerland, and Polypeptide, Denmark.

Proteins are "very HMW" (MW > 100,000) organic compounds, consisting of amino acid sequences linked by peptide bonds. They are arranged in linear chains and folded into globular forms. They are essential to the structure and function of all living cells and viruses and are among the most actively studied molecules in biochemistry. They can be made only by advanced biotechnological processes, primarily mammalian cell cultures (see Section 4.2). Monoclonal antibodies (mAb) prevail among human-made proteins. About a dozen of them are approved as pharmaceuticals. Important modern products are EPO (Binocrit, NeoRecormon, erythropoietin), sales $3.5 billion (2009); Enbrel (etanercerpt), $3.5 billion; Remicade (infliximab), $4.4 billion; MabThera/Rituxin (rituximab), $5.9 billion; and Herceptin (trastuzumab), $5.1 billion.

PEGylation is a big step forward regarding administration of peptide and protein drugs. The method offers the two-fold advantage of substituting injection by oral administration and reducing the dosage, and therefore the cost of the treatment. The pioneer company in this field is Prolong Pharmaceuticals which has developed a PEGylated erythropoietin (PEG-EPO).

Oligonucleotides are a third category of big molecules. They are oligomers of nucleotides, which in turn are composed of a five-carbon sugar (either ribose or desoxyribose), a nitrogenous base (either a pyrimidine or a purine), and 1–3 phosphate groups. The best-known representative of a nucleotide is the coenzyme ATP(=adenosine-triphosphate, MW 507.2). The structure of ATP consists of adenine, attached at the 1′ carbon of ribose and three phosphate groups attached at the 5′ position:

Oligonucleotides are chemically synthesized from protected phosporamidites of natural or chemically modified nucleosides. The oligonucleotide chain assembly proceeds in the direction from 3′- to 5′-terminus by following a procedure referred to as a "synthetic cycle." Completion of a single synthetic cycle results in the addition of one nucleotide residue to the growing chain. The maximum length of synthetic oligonucleotides hardly exceeds 200 nucleotide components.

From its current range of applications in basic research, a.o. in drug target validation, drug discovery and therapeutic development, the potential use of oligonucleotides is foreseen in gene therapy (antisense drugs), disease prevention, and agriculture.

Peptides and *oligonucleotides* are now often summarized under the heading "tides." They are used in a variety of pharmaceutical applications, including antisense agents that inhibit undesirable cellular protein production, which causes many diseases; as antiviral agents; and as protein binding agents. An antisense drug in advanced (Phase III) development is Genzyme's cholesterol lowering drug Mipomersen.

Antibody-drug conjugates (ADC) constitute a combination between small and big molecules. The small-molecule part, up to four different APIs, are highly potent cytotoxic drugs. They are linked with a monoclonal antibody, a big molecule which is of little or no therapeutic value in itself, but extremely discriminating for its targets, the cancer cells. The first commercialized ADCs were Isis's Formivirisen and, more recently, Pfizer's (*formerly* Wyeth) Mylotarg (gemtuzumab ozogamicin), a conjugate of N-acetyl-γ-calicheamicin with the humanized mouse monoclonal IgG4 κ antibody, hP67.6. Examples of ADCs in Phase III of development are Abbott's/Isis's Alicaforsen and Eli Lilly's Aprinocarsen.

CHAPTER 4

Technologies

Several key technologies are used for the production of fine chemicals, including

- Chemical synthesis, either from petrochemical starting material or from extracts from natural products.

  > Biological sources—there is an estimate of 10–100 million different forms of life on earth—are still only scarcely investigated: For instance, out of an estimated number of 1.5 million fungi, only 70,000 are known; and out of an estimated 0.4–3 million bacteria, only 6000 are known.

- Biotechnology, in particular biocatalysis (enzymatic methods), fermentation, and cell culture technology.
- Extraction from animals, microorganisms, or plants; isolation and purification, used, for example, for alkaloids, antibacterials (especially penicillins), and steroids.
- Hydrolysis of proteins, especially when combined with ion exchange chromatography, used, for instance, for amino acids.

Chemical synthesis and biotechnology are most frequently used; sometimes also in combination. They are described in the following sections.

4.1 TRADITIONAL CHEMICAL SYNTHESIS

Two general methods, the "bottom-up" and "top-down" approaches, are used for synthesizing fine chemicals (see Section 6.1).

Examples of the reactions used to synthesize a number of well-known pharmaceuticals are shown in Table 4.1. The number of synthetic steps required to make the desired active pharmaceutical ingredients (APIs) ranges from two (acetaminophen) to seven (omeprazole).

Fine Chemicals: The Industry and the Business, Second Edition, by Peter Pollak
Copyright © 2011 John Wiley & Sons, Inc.

TABLE 4.1   Reactions Used to Synthesize Selected APIs

API		Reactions
Name	Formula	
Acetaminophen (Paracetamol)	N-(4-hydroxyphenyl) acetamide	Partial hydrogenation of nitrobenzene, N-acetylation
Amoxicillin	6-[D-)-α-Amino-p-hydroxyphenyl-acetamido] penicillanic acid	Aromatic alkylation, amination, imine formation, amidation (side chain), fermentation, and deamidation (penicillin nucleus)
Cephalexin	7-(D-α-Aminophenyl-acetamido) cephalosporanic acid	Aromatic alkylation, amination, imine formation, amidation (side chain) Fermentation, deamidation (penicillin nucleus), acid-catalyzed ring expansion
Guaifenesin	Glycerol mono (2-methoxy-phenyl) ether	Peracid oxidation of henol, partial N-methylation,nucleophilic displacement of glycidol
Ibuprofen	(±)-2-(4-Isobutylphenyl) propionic acid	Aromatic alkylation, HF-catalyzed aromatic acetylation, palladium-catlayzed carbonylation, alkene hydration
Lisinopril	(S)-1-N^2-(1-Carboxy-3-phenylpropyl)-L-lysyl] L-proline dihydrate	Pd-catalyzed carbonylation, hydroxylation of a double-bond, chemical rcsolution, amidation, N-protection (using trifluoroacetic anhydride), carbonyl activation (using phosgene)
Omeprazole	5-Methoxy-2-{[(4-methoxy-3,5-dimethyl-2-pyridinyl) methyl] sulfinyl}-1H-benzimidazol	O-methylation, imidazole ring formation using thiourea, N-oxidation, nucleophilic displacement, N-methylation, S-oxidation
Sulfamethoxazole	4-Amino-N-(5-methyl-3-isoxazolyl) benzene sulfonamide	Sulphonylation of aniline, sulfoamidation with 3-amino-5-methyl-isoxazol

Source: Adapted from Brychem.

   For each step of a synthesis, a large toolbox of chemistries is available. Most of them have been developed on laboratory scale by academia over the last century and subsequently adapted to industrial scale. For example, "evaporating to dryness" had to be elaborated to "concentrate in a thin-film evaporator and precipitate by addition of isopropanol." The two most comprehensive handbooks describing organic synthetic methods are the *Encyclopedia of Reagents for Organic Synthesis* [1] and *Houben-Weyl, Methods of Organic Chemistry* [2][1].

More than 150 types of reaction offered by the fine chemical industry are listed in the process directory section of the *Informex Show Guide* [3]; 45 of them are organic name reactions, representing 10% of a comprehensive listing in the *Merck Index*. They range from acetoacetylation all the way through to Wittig reactions. Each of the 430 companies participating at the survey indicated competence for close to 30 types of reaction on average. Amination, condensation, esterification, Friedel–Crafts, Grignard, halogenation (especially chlorination), hydrogenation, and reduction, respectively (both catalytic and chemical) are most frequently mentioned. Optically active cyanohydrin, cyclopolymerization, ionic liquids, nitrones, oligonucletides, peptide (both liquid- and solid-phase), electrochemical reactions (e.g., perfluorination) and steroid synthesis are promoted by only a limited number of companies. With the exception of some stereospecific reactions, particularly biotechnology (see Section 4.2), mastering these technologies does not represent a distinct competitive advantage. Most reactions can be carried out in standard multipurpose plants. The very versatile organometallic reactions (e.g., conversions with lithium aluminum hydride, boronic acids) may require temperatures as low as −100°C, which can be achieved only in special cryogenic reaction units, either by using liquefied nitrogen as coolant or by installing a low-temperature unit. Other reaction-specific equipment, such as filters for the separation of catalysts, ozone, or phosgene generators, can be purchased in many different sizes. The installation of special equipment generally is not a critical path on the overall project for developing an industrial-scale process of a new molecule.

Since the mid-1990s the commercial importance of single-enantiomer fine chemicals has increased steadily. They constitute about half of both existing and developmental drug APIs. In this context, the ability to synthesize chiral molecules has become an important competency. There are basically two types of processes available, namely the traditional physical separation of the enantiomers and a stereospecific synthesis, using chiral catalysts. Among the chiral catalysts, enzymes and synthetic BINAP types (see Table 11.11) are used most frequently. Large volume (>10^3 mtpa) processes using chiral catalysts include the manufacture of the perfume ingredient l-Menthol and Syngenta's Dual (metolachlor) as well as BASF's Outlook (dimethenamid-P) herbicides. Examples of originator drugs, which apply asymmetric technology, are AstraZeneca's Nexium (esomeprazole), which uses chiral oxidation, and Merck's Januvia (sitagliptin), where an asymmetric hydrogenation of an unprotected enamine is carried out in the almost final step of the synthesis.

The physical separation of chiral mixtures and purification of the desired enantiomer can be achieved either by classical crystallization (having a

---

[1] Bracketed reference numbers cited in text throughout the book correspond to "Cited Publications" entries in Bibliography sections at the end of Parts I–III.

"low-tech" image but is still widely used), using standard multipurpose equipment or by various types of chromatographical separation, such as standard column, simulated moving bed (SMB), or supercritical fluid (SCF) techniques. The latter are advanced chromatographic technology for the separation of demanding racemates and elimination of trace impurities.

With the exception of some stereospecific reactions, particularly biotechnology (see Section 4.2), mastering these technologies does not represent a distinct competitive advantage. Most reactions can be carried out in standard multipurpose plants. The very versatile organometallic reactions (e.g., conversions with lithium aluminum hydride, boronic acids) may require temperatures as low as −100°C, which can be achieved only in special cryogenic reaction units, either by using liquefied nitrogen as coolant or by installing a low-temperature unit. Other reaction-specific equipment, such as ozone or phosgene generators, can be purchased in many different sizes. The installation of special equipment generally is not a critical path on the overall project for developing an industrial-scale process of a new molecule.

Microreactor Technology (MRT), making part of "process intensification," is a relatively new tool that is being developed at several universities,[2] as well as leading fine chemical companies, such as Bayer Technology Services, Germany; Clariant, Switzerland; Evonik-Degussa, Germany; DSM, The Netherlands; Lonza, Switzerland; PCAS, France, and Sigma-Aldrich, USA. The latter company produces about 50 fine chemicals up to multi-kilogram quantities in microreactors. From a technological point of view, MRT represents the first breakthrough development in reactor design since the introduction of the stirred-tank reactor, which was used by Perkin & Sons, when they set up a factory on the banks of what was then the Grand Junction Canal in London in 1857 to produce mauveïne, the first-ever synthetic purple dye. For a comprehensive coverage of the subject, see reference 4.

The main advantages of microreactors, aka continuous flow reactors, are (1) much better heat and mass transfer, allowing one to carry out energetic reactions safely and rapidly, and obtain higher yields, selectivities, and product quality and (2) the substantially shortened development times. Processes that are run in microreactors do not need the cumbersome scale-up from laboratory to pilot plant to industrial-scale plant. Capacity increases are achieved both by "scale-out" of module volume and "numbering up," namely using more units in parallel. Also regulatory issues are becoming much simpler. FDA's Process Analytical Technology (PAT) initiative encourages pharmaceutical fine chemical (PFC) producers to use continuous processes wherever possible. Disadvantages are the high investment cost, problems associated with solids handling (already the precipitation of a by-product can cause clogging), the difficulty to precisely assign feed streams for multiple units, to clean the

---

[2] A.o. Swiss Federal Institute of Technology, Zurich (ETHZ), Switzerland; Massachusetts Institute of Technology (MIT), USA; Institut für Mikrotechnik (IMM), Germany; University of Washington and Micro-Chemical Process Technology Research Association (MCPT), Japan.

equipment following current Good Manufacturing Practice (cGMP) procedures, and—last but not least—the lack of experience with "industrial-scale" production over extended periods of time. Examples for reactions that have worked in microreactors include aromatics oxidations, diazomethane conversions, Grignards, halogenations, hydrogenations, nitrations, and Suzuki couplings. Japanese companies, with their heritage in fine arts and crafts, have assumed a leadership position in microtechnology.

According to experts in the field, 70% of all chemical reactions could be done in microreactors; however, only 10–15% are economically justified.

The first industrial-scale use of MRT for producing a PFC under cGMP conditions is for Naproxcinod. This 4-nitrobutylester homolog of naproxen is a promising nonsteroidal anti-inflammatory drug under advanced development by the French virtual pharma company NiCox. The crucial step of the synthesis is the highly energetic, selective mono-nitration of butane-1,4-diol. The production asset of the custom manufacturer, DSM, consists of a cluster of microreactors, continuous extraction columns, centrifugal extractors, and distillation equipment. The name plate annual production capacity is 16 tons [3].

Psychological, technical, logistic, and financial challenges must be resolved, before a widespread use of MRT in the fine chemical industry will become a reality. Whenever a chemical synthesis is studied in a university or industrial laboratory, the chemist uses a stirred glass flask, that is, a batch reactor ... and must change his mind-set in order to "act continuously." A technical challenge, apart from the abovementioned clogging caused by solids formation, is the requirement for a uniform feed of multiple reactors. On the logistics front, the production planning must cope with syntheses steps 1 and 2, 4 and 5 of an API carried out in batch reactors with a capacity of, say, several tons per day each, and step 3, carried out in a microreactor with a capacity of less than 100 kg/day. Last but not least, it will be difficult to justify an investment in an MRT production line, which per se is expensive in terms of dollars/cubic meter × hour capacity, when an underutilized, depreciated batch production line is available!

For *peptides*, three main types of methods are used, namely chemical synthesis, extraction from natural substances, and biosynthesis. Chemical synthesis is used for smaller peptides made of up to 30–40 amino acids. One distinguishes between "liquid phase" and "solid phase" synthesis. In the latter, reagents are incorporated in a resin that is contained in a reactor or column. The synthesis sequence starts by attaching the first amino acid to the reactive group of the resin and then adding the remaining amino acids one after the other. In order to ascertain a full selectivity, the amino groups have to be protected in advance. Most developmental peptides are synthesized by this method, which lends itself to automation. By developing both more efficient protecting groups and resins, it has been improved substantially over the last 50 years. As the intermediate products resulting from individual synthetic steps cannot be purified, a virtually 100% selectivity is essential for the synthesis of larger-peptide molecules. Even at a selectivity of 99% per reaction step, the purity will drop to less than 75% for a dekapeptide (30 steps)! Therefore, for industrial quantities of peptides, not more than 10–15 amino acid peptides can be made using the solid-phase method. For laboratory quantities, up to 40 are possible. In order to prepare larger peptides, individual fragments are first produced, purified, and then combined with the final molecule by liquid phase synthesis. This combination of methods is listed under "chemical hybrid" in Table 4.2. Thus, for the production of Roche's anti-AIDS drug Fuzeon

TABLE 4.2   Comparison of Peptide Manufacturing Technologies

Method	Advantages	Limitations
Chemical synthesis		
Chemical— solution phase	Potential for scale-up to metric tonnes	Lengthy and costly development, relatively high cost of raw materials and conversion
Chemical—solid phase	Rapid development for small to medium scale	Scale-up potential may be limited; relatively high cost of raw materials and conversion
Chemical— hybrid	Relatively rapid development cycle; potential for scale-up to metric tonnes	Raw materials (resins, amino acid derivatives) currently expensive
Extraction	Relatively simple scale-up	Scale limited by availability of source; potential for contamination from source material
Fermentation	Low raw material costs; virtually unlimited scale-up	Only applicable for naturally occurring products
Semi-synthesis	Potential for complex molecules containing unnatural amino acids	Source availability
Enzymatic	Mild coupling under aqueous conditions without the need for side-chain protection	Competing proteolysis limits usefulness
Recombinant— fermentation	Low raw material costs; scale-up potential limited only by size of production facility; purification relatively easy.	Relatively costly development; unnatural sequences may be difficult or impossible
Recombinant— transgenic animals and plants	Low raw material costs; unlimited scale-up; purification may be relatively easy	Costly development and production; unnatural sequences not possible

Source: PolyPeptide.

(enfuvirtide), three fragments of 10–12 amino acids are first made by solid-phase synthesis and then linked together by liquid-phase synthesis. The preparation of the whole 35 amino acid peptide requires more than 130 individual steps! Most developmental peptides are synthesized by this method, which lends itself to automation.

> "Solid-phase" peptide synthesis was pioneered by R. B. Merrifield who introduced it in the early 1960s. By developing both more efficient protecting groups and resins, it has been improved substantially over the last 50 years. Nowadays, the leading solid phases are the "2-chlorotrityl chloride resins." They consist of a polystyrene-base resin cross-linked with a small amount of divinyl-benzene and functionalized with 2-chlorotritiyl chloride. Professor K. Barlos (University of Patras, Greece) has made the single most important contribution to the development of these resins.

Whereas the overall demand for outsourced PFCs is expected to increase moderately (see Chapter 15), the estimated annual growth rates for the above-mentioned niche technologies are much higher. Microreactors and the SMB separation technology are expected to grow at a rate of even 50–100% per year! It has to be realized, however, that the total size of the accessible market typically does not exceed a few hundred tons per year at best.

4.2 BIOTECHNOLOGY

Industrial biotechnology, also called "white biotechnology,[3]" is increasingly impacting the chemical industry, enabling both the conversion of renewable resources, such as sugar or vegetable oils, and the more efficient transformation of conventional raw materials into a wide range of commodities, such as ethanol, fine chemicals (e.g., 6-aminopenicillanic acid), and specialties (e.g., food and feed additives). As opposed to green and red biotechnology, which relates to agriculture and medicine, respectively, white biotechnology enables the production of existing products in a more economic and sustainable fashion on the one hand, and provides access to new products, especially biopharmaceuticals, on the other hand. Three very different process technologies—biocatalysis, biosynthesis (microbial fermentation), and cell cultures—are used.

Biocatalysis

Biocatalysis, also termed *biotransformation* and *bioconversion*, makes use of natural or modified isolated enzymes, enzyme extracts, or whole cell systems for enhancing the production of small molecules. It has much to offer

---

[3] Green biotechnology is applied to agricultural processes, for example, for designing genetically modified plants. Red biotechnology is applied to medical processes, such as engineering of genetic cures through genomic manipulation.

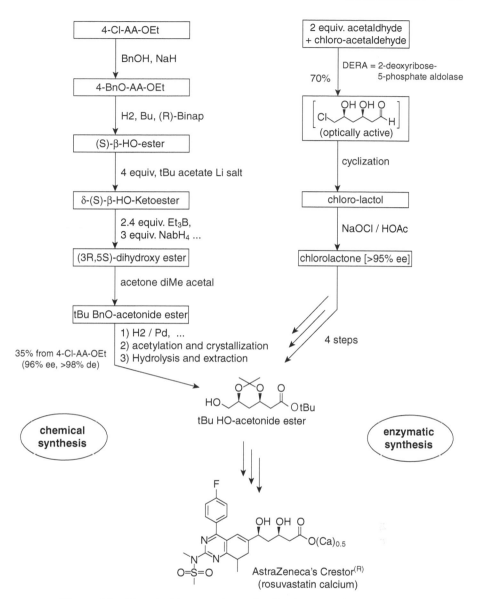

Figure 4.1   Chemical/enzymatic synthesis of Crestor (rosuvastatin).

compared with traditional organic synthesis. It involves mild reaction conditions (ambient temperature and pressure at physiological pH) and affords
high chemo-, regio-, and stereoselectivities. Furthermore, it generally requires
fewer steps, by eliminating the need for protection and deprotection steps
(see Figs. 4.1 and 4.2) and avoids the use of environmentally unattractive
organic solvents. The syntheses are shorter, less energy intensive and generate

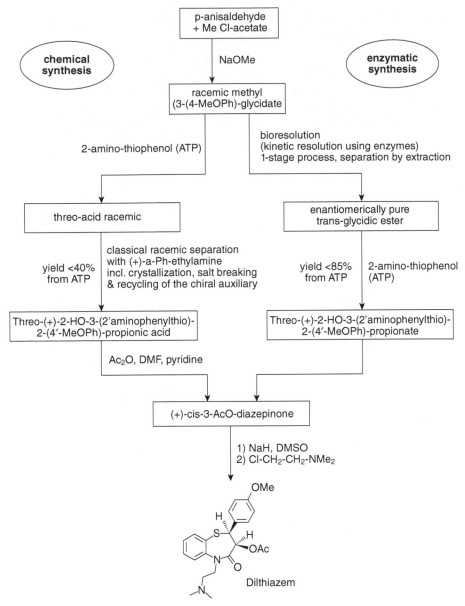

Figure 4.2   Chemical/enzymatic synthesis of dilthiazem.

less waste, and hence, are both environmentally and economically more attractive. A starting material is converted by the biocatalyst in the desired product. Enzymes are differentiated from chemical catalysts particularly with regard to stereoselectivity, regioselectivity, and chemoselectivity. Whereas enzymes were traditionally associated with the metabolic pathway of natural substances, they can also be tailored for use in chemical synthesis. Biocatalysts, particularly enzymes, are applied like chemical catalysts, either in solution or on solid supports ("immobilized enzymes"), albeit under milder conditions and in aqueous solution. Immobilized enzymes can be recovered by filtration after completion of the reaction. Conventional plant equipment can be used without, or only modest, adaptations.

The International Union of Biochemistry and Molecular Biology has developed a classification for enzymes. The main categories are

- *Oxidoreductases*, which catalyze oxidation–reduction reactions and are acting, for example, on aldehyde or keto groups. An important application is the synthesis of chiral molecules, especially chiral PFCs (about two-thirds of chiral products produced on large industrial scale are already made using biocatalysis).
- *Transferases*, which transfer a functional group, for example, $-CH_3$ or $-OPO_3$.
- *Hydrolases*, which catalyze the *hydrolysis* of various bonds. The best-known subcategory of hydrolases are the lipases, which hydrolyze ester bonds. In the example of human pancreatic lipase, which is the main enzyme responsible for breaking down fats in the human digestive system, a lipase acts to convert triglyceride substrates found in oils from food to monoglycerides and free fatty acids. In the chemical industry, lipases are also used, for instance, to catalyze the $-C{\equiv}N \rightarrow -CONH_2$ reaction, for the synthesis of acrylamide from acrylonitril, or nicotinic acid from 3-pyridylnitrile.
- *Lyases*, which cleave various bonds by means other than hydrolysis and oxidation, such as starch to glucose.
- *Isomerases*, which catalyze isomerization changes within a single molecule.
- *Ligases*, which join two molecules with covalent bonds.

Whereas in the past, only about 150 out of 3000 known enzymes were used commercially, new developments in technology are increasing this number dramatically. Both natural diversity and synthetic "reshuffling" are being more and more exploited to obtain enzymes from diverse environments and with a large variation in properties. Companies specializing in making enzymes, such as Novozymes, Danisco (Genencor), or modifying (respectively "tailoring") enzymes to specific chemical reactions, such as Codexis, have yielded enormous progress regarding areas of application, specificity, concentration, throughput, stability, ease of use, and economics. Nonetheless, the commercialization of many enzymatic processes is hampered by the lack of operational stability of many enzymes, coupled with their relatively high price.

The highest-volume chemicals made by biocatalysis are bioethanol (70 million MT), high-fructose corn syrup (2 million MT), acrylamide, 6-amino-penicillanic acid (APA), L-lysine and other amino acids, citric acid, and niacinamide (all more than 10,000 MT). Examples of modern drugs, where enzymes are used in the synthesis, are Pfizer's Lipitor (atorvastatin), where the pivotal intermediate R-3-Hydroxy-4-cyanobutyrate is now made with a nitrilase, and Merck's Singulair (montelukast), where the reduction of a ketone to S-alcohol, which had required stoichiometric amounts of expensive and moisture sensitive "(–)-DIP chloride" is now replaced by a ketoreductase enzyme catalyst step.

In the manufacture of fine chemicals, enzymes represent the single most important technology for radical cost reductions. This is particularly the case in the synthesis of molecules with chiral centers. Here, it is possible to substitute the formation of a salt with a chiral compound, for example, (+)-α-phenylethylamine, crystallization, salt breaking, and recycling of the chiral auxiliary, resulting in a theoretical yield of not more than 50%, with a one-step, high-yield reaction under mild conditions and resulting in a product with a very high enantiomeric excess (ee). Two prime examples are AstraZeneca's blockbuster drug Crestor (rosuvastatin; see Fig. 4.1) and the widely used generic dilthiazem (see Fig. 4.2), and a Biocatalysis pilot plant, Degussa-Evonik.

Although the original synthesis already had been improved by substituting the cumbersome racemate separation (see above) by a stereo-specific catalytical step, it still presented serious deficiencies due to the physical properties of certain intermediates, such as high boiling points of temperature-sensitive oily substances requiring an ultra-HV distillation, the cost of some raw materials (notably 4 equivalents tBu acetate Li salt and boranes), and the overall length of the process. By switching to the enzymatic route, which has been developed by Chi-Huey Wong, a chemistry professor at Scripps Research Institute in 1994, a 70% cost reduction has been achieved for the pivotal intermediated, tBu HO-acetonide ester.

The enzymatic process for dilthiazem is another landmark. It allowed the reduction of the cost of goods sold (COGS) for this anti-anginal, antihypertensive, and anti-arrhythmic drug by 40%. The two main advantages of the process, which was developed by DSM, are an operational simplification (much smaller plant, 10 times higher throughput) and cost savings on raw materials, mainly 2-amino-thiophenol, which is used on a later stage in the process.

Similar rewarding switches from chemical steps to enzymatic ones have also been achieved in steroid synthesis. Thus, it has been possible to reduce the number of steps required for the synthesis of dexamethasone from bile from 28 to 15—and further reductions are in the making!

Biosynthesis

Biosynthesis by microbial fermentation, that is, the conversion of organic materials into fine chemicals by microorganisms, is used for the production of both small molecules (using enzymes in whole cell systems) and less complex, non-glycosylated big molecules, including peptides and simpler proteins. The technology has been used for 10,000 years to produce food products, such as alcoholic beverages, cheese, yogurt, and vinegar. In contrast to biocatalysis, a biosynthetic process does not depend on chemicals as starting materials, but only on cheap natural feedstock, such as glucose, to serve as nutrient for the cells. The enzyme systems triggered in the particular microorganism strain lead to the excretion of the desired product into the medium or, in the case of high-molecular-weight (HMW) peptides and proteins, to the accumulation within so-called inclusion bodies in the cells. The key elements of fermentation development are strain selection and optimization, media, and process development. For the large-scale industrial production of fine chemicals and proteins, dedicated plants are used. As the volume productivity is low, the bioreactors, called *fermenters*, are large, with volumes that can exceed $250\,m^3$. Product isolation was previously based on large-volume extraction of the medium containing the product. Modern isolation and membrane technologies, such as reverse osmosis, ultra- and nano-filtration, or affinity chromatographic methods can help to remove salts and by-products and to concentrate the solution efficiently and in an environmentally friendly manner under mild conditions. The final purification is often achieved by conventional chemical crystallization processes.

In contrast to the isolation of small molecules, the isolation and purification of microbial proteins is tedious and often involves a number of expensive large-scale chromatographic operations.

Examples of large-volume LMW products made by modern industrial microbial biosynthetic processes are monosodium glutamate (MSG), vitamin B_2 (riboflavin), and vitamin C (ascorbic acid). In vitamin B_2, riboflavin, the original six- to eight-step synthetic process starting from barbituric acid, has been substituted completely by a microbial one-step process, allowing a 95% waste reduction and an approximately 50% manufacturing cost reduction. In ascorbic acid, the five-step process (yield ≈85%) starting from D-glucose, originally invented by Tadeus Reichstein in 1933, is being gradually substituted by a more straightforward fermentative process with 2-ketogluconic acid as pivotal intermediate. After the discovery of penicillin in 1928 by Sir Alexander Fleming from colonies of the bacterium *Staphylococcus aureus*, it took more than a decade before Howad Florey and Ernst Chain isolated the active ingredient and developed a powdery form of the medicine. Since then, many more antibiotics and other secondary metabolites have been isolated and manufactured by microbial fermentation on a large scale. Some important antibiotics besides penicillin are cephalosporins, azithromycin, bacitracin, gentamicin, rifamycin, streptomycin, tetracycline, and vancomycin.

More recently, GlaxoSmithKline patented an efficient fermentation route for the biosynthetic production of thymidine (thymine-2-desoxyriboside). Key to the invention is a recombinant strain that efficiently produces high titers of thymidine by blocking some enzymes in the thymidine regulating pathway. This microbial process has now replaced the chemical route and has enabled GSK to supply the anti-AIDS drug AZT (zidovudine) to third-world countries at low cost.

Cell Cultures

In mammalian organisms, the glycosylation of proteins, involving the addition of sugar moieties, often is essential for biological activity. The ability of a particular cell or organism to correctly glycosylate a protein can determine its usefulness to make a given protein. Animal or plant cells, removed from tissues, will continue to grow if supplied with and under the appropriate nutrients and conditions. When carried out outside the natural habitat, the process is called *cell culture*. Mammalian cell culture fermentation is used mainly for the production of complex big molecules with specific glycosylation patterns and tertiary protein structures, such as therapeutic proteins and monoclonal antibodies (mAbs). Typical biotech APIs obtainable through mammalian cell culture are: viral vaccines, hormones, immunobiologicals (such as interleukin, lymphokines, mAbs), blood clot dissolvers (tPAs: tissue plasminogen activators), and a number of complex glycosylated proteins

Related technologies use plant cell cultures, insect cells, or transgenic animals. In contrast to the mammals, bacteria, plants, and fungi are incapable of glycosylation. Cell culture processes allow single cells to act as independent units, much like a microorganism such as a bacterium or fungus. The cells are capable of dividing; they increase in size and, in a *batch culture*, can continue to grow until limited by some culture variable such as nutrient depletion.

Mammalian cell culture, also known as *recombinant DNA technology*, has existed for 50 years. It serves for producing HMW, or simpler, big-molecule fine chemicals, including glycoproteins and mAbs. The first products made were interferon (discovered in 1957), insulin, and somatropin (see Section 3.2). For mammalian cell culture, specific cell lines are developed. It is a uniform cell population that can be cultured continuously. Commonly used cell lines are Chinese hamster ovary (CHO) cells or plant cell cultures (see text below). The production volumes are very small. The demand exceeds 100 kg per year for only four products: Rituxan (Roche-Genentech), Enbrel (Amgen and Merck [*formerly* Wyeth]), and Remicade (Johnson & Johnson [J&J]). The need for cell culture technology stems mainly from the fact that bacteria do not have the ability to perform many of the posttranslational modifications that most large proteins require for *in vivo* biological activity. These modifications include intracellular processing steps such as protein folding, disulfide linkages, glycosylation, and carboxylation. Also, mammalian cell culture is carried out using bioreactor batches, but industry convention states that the term "fermentation" is correctly used when handling with unicellular organisms (such as bacteria or yeasts), whereas the term "cell culture" must be used for bioreactors handling with multicellular organisms (including, thereby, mammalian cells [a.o CHO]).

Mammalian cell culture, however, is a delicate operation, posing more problems than handling with bacteria. The bioreactor batch requires more stringent controls of operating parameters, since mammalian cells are heat and share sensitive; in addition, the growth rate of mammalian cells is very slow, requiring from days to several months. Thereby, due to the high costs of growing mammalian cells, this technology is used only when strictly indispensable. While there are substantial differences between microbial and mammalian technologies (e.g., the volume/value relationships are $10/kg and 100 tons for microbial, $1,000,000/kg and 10 kg for mammalian technology; the cycle times are 2–4 and 10–20 days, respectively), they are even more pronounced between mammalian and synthetic chemical technology (Table 4.3).

TABLE 4.3 Key Characteristics of Biotechnological and Chemical API Manufacturing (All Figures Are Indicative Only)

	Mammalian Cell Technology	Chemical Technology
Worldwide reactor volume	≈3000 m³ (Fermenters)	≈ 80,000 m³
Investment per cubic meter reactor volume	≈$5 million[a]	≈$500,000
Production per cubic meter reactor volume and year	Several 10 kg	Several 1000 kg
Sales per cubic meter reactor volume and year	≈$5–10 million	≈$250,000–500,000
Value of 1 batch	≈$5 million (20,000-L Fermenter)	≈$500,000
Product concentration in reaction mixture	≈2–6 (–10)g/Liter	≈100 g/Liter (10%)
Typical reaction time	≈20 days	≈6 h
Process development time	≈3 years (one step)	2–3 months per step
Capacity expansion projects	Many, doubling of actual capacity	few, mainly in Far East
Governing rules	cGMP, BLA[b]	cGMP, ISO 14000
Scale-up factor (1st lab process to industrial scale)	≈10^9 (μg → 1 ton)	≈10^6 (10 g → 10 tons)
Process development time	≈3 years (one step)	2–3 months per step
Plant construction time	4–6 years	2–3 years
Share of outsourcing		
Early stage	55%	25% of chemical prod.
Commercial	20%	45% of chemical prod.

[a] Examples: *Boehringer-Ingelheim* invested $320 million, resp. $3.5 million/m³ bioreactor volume, in a new 90,000-L eight-floor biopharmaceutical production plant at the former Dr. K. Thomæ site in Biberach/Riss, Germany.

Lonza invested $275 million for addition of 60,000-L fermenter volume in Portsmouth, NH, corresponding to = $4.58 million/m³).

Merck Serono plans to invest €300 million for addition of 120,000-L fermenter volume by 2012 in Corsier-sur-Vevey, Switzerland, corresponding to $3.75 million/m³.

[b] Biological License Application (product specific).

Source: Reference [5].

Typical production volumes for biopharmaceuticals made by mammalian cell technology are EPO-alpha, 7 kg (worldwide); Etanercept, 463 kg (USA), Rituximab, 418 kg (USA), and Adalimumab, 61 kg (USA).

The low productivity of the animal culture makes it very vulnerable to contamination, as a small number of bacteria would soon outgrow a larger population of animal cells. Given the fundamental differences between the two process technologies, plants for mammalian cell culture technologies have to be built *ex novo*. The production of biopharmaceuticals starts by inoculating a nutrient solution with cells from a cell bank. The latter are allowed to reproduce in stages on a scale of up to several thousand liters. The cells secrete the desired product, which is then isolated from the solution, purified, and transferred to containers. A process flow sheet for protein production from mammalian cells is shown in photo 8 in the insert.

The mammalian cell production process is divided into the following four main steps:

1. *Cultivation.* The cells are transferred from the cryogenic cell bank to a liquid nutrient medium, where they are allowed to reproduce. Mammalian cells such as CHO divide about once every 24 h (bacterial cells, such as *Escherichia coli*, usually divide once every 20 min, and thus a sufficient number of cells are obtained in a much shorter time than in traditional fermentation processes). During the growth phase, the cell culture is transferred to progressively larger culture vessels.

2. *Fermentation.* The actual production of the biopharmaceutical occurs during this phase. The culture medium contains substances needed for synthesis of the desired therapeutic protein. In total, the medium contains around 80 different constituents at this stage, although manufacturers never disclose the exact composition. The industrial-scale bioreactors have capacities of 10,000 L or multiples. There are both technological and biological constraints on the size of the reactor—the bigger a fermenter is, the more difficult it becomes to create uniform conditions around all the cells contained in it. The fermentation reaction is done in a batch, fed-batch, or perfusion mode (see, e.g., Airlift fermentor, fermentation pilot plant, Lonza, Slough, UK and its Biologics plant,, in Portsmouth NH, USA).

3. *Purification.* The production of biopharmaceuticals in cells is a one-step process, and the product can be purified immediately after fermentation. In the simplest case, the cultured cells will have secreted the product into the ambient solution. Thus, the cells are separated from the culture medium (e.g., by centrifugation or filtration), and the desired product is then isolated via several purification, typically chromatographic steps. If, on the other hand, the product remains in the cells following biosynthesis, the cells are first isolated and digested (i.e., destroyed), and the cellular debris is then separated from the solution together with

the product. As both the reaction times are long and the product concentration is small, the productivity of this technology is low. For example, a 10,000-L fermenter yields only a few kilograms of a therapeutic antibody, such as rituximab or trastuzumab. The production steps, including purification, take several weeks. Several more weeks are then needed to test the product. Each product batch is tested for purity to avoid quality fluctuations, and a 99.9% purity level is required for regulatory approval. Only then can the finished product be formulated and shipped.

4. *Formulation.* The final steps in the production of biopharmaceuticals are also demanding. The sensitive proteins are converted to a stable pharmaceutical form and must be safely packaged, stored, transported, and finally administered. Throughout all these steps, the structural integrity of the molecule has to be safeguarded to maintain efficacy. At present, this is possible only in special solutions in which the product can be cryogenically frozen and preserved, although the need for low temperatures does not exactly facilitate transport and delivery. Biopharmaceuticals are therefore strictly made to order.

Because of the sensitive nature of most biopharmaceuticals, their dosage forms are limited to injectable solutions. Therapeutic proteins cannot pass the acidic milieu of the stomach undamaged, nor are they absorbed through the intestinal wall. Although work on alternatives such as inhalers is underway (the first commercial application is an insulin inhaler), injection remains the predominant option for administering sensible biopharmaceuticals.

Nowadays, all the steps in the production of biopharmaceuticals are fully automated. Production staff steps in only if problems occur. Even a trace amount of impurities can cause considerable economic loss, as the entire production batch then has to be discarded, the equipment dismantled and cleaned, and the production process restarted from scratch with the cultivation of new cells.

Plant cell culture is in an early stage of technology development. Plants produce a wide range of secondary metabolites, some of which have been found to be pharmacologically active. However, these compounds are generally produced in very small amounts over a long period of time, making commercially viable extraction difficult. The technology shows promise for the selective synthesis of chiral compounds with a polycyclic structure, as found in many cytostatics, such as camptothecin, vinblastine, and paclitaxel (see Section 15.2). The concentration of biologically active molecules within the plant is usually very low. Apart from pure manufacturing and weather-related factors in manufacturing pharmaceutical substances in living plants, the downstream processing, isolation, and purification technologies that need to be developed are key to the overall process costs. The first API which is about to demonstrate the industrial viability and economy of scale is taliglucerase alfa. It is a drug for the treatment of Gaucher's disease in advanced development at Protalix, Israel. It is derived from a proprietary plant cell-based expression platform using genetically engineered carrot cells.

The "Deutsche Sammlung von Mikroorganismen und Zellkulturen GmbH—DSMZ" (German Collection of Microorganisms and Cell Cultures) is the most comprehensive Biological Resource Centre in Europe. The nonprofit organization counts more than 18,000 microorganisms, 1200 plant viruses, 600 human and animal cell lines, 770 plant cell cultures, and more than 7100 cultures deposited for the purpose of patenting.

CHAPTER 5

Facilities and Plants

5.1 PLANT DESIGN

Fine chemical plants are either located on a separate industrial site or part of a larger chemical complex. A typical stand-alone site comprises one or more production buildings, an administrative building, laboratories, a warehouse (with separate sections for raw materials, quarantine products, and finished products), a power station for steam generation, a utilities building (for demineralized water, steam, brine, inert gas, etc. preparation), waste treatment facilities (for the treatment of organic and aqueous liquid and solid waste and off gases), a maintenance shop, a tank farm for liquid raw materials and solvents and, in dry areas, a cooling tower. Of these facilities, only the production building is fine chemical specific. It typically consists of three distinct sections: a reaction part, also referred to as "wet section," a product finishing part, also referred to as "dry section," and an administrative part, comprising quality control laboratories, offices, change rooms, and other services.

The basic type of plant, batch or continuous, is strongly correlated to the production volume of the products made. Continuous plants are used for the production of all top 300 commodities in terms of production volume. For fine chemicals, which *per definitionem* are produced in smaller quantities, however, the picture is entirely different. For those ranking 300 to 3000, 90%, and for those above 3000, 97% are made in batch-type plants. Particularly in custom manufacturing, it does not make sense to build dedicated production units for individual products, because the requirement for single fine chemicals rarely exceeds 100 tons/year (see Fig. 5.1), and because the majority of fine chemicals can be produced in standardized equipment. Moreover, the product portfolio is regenerated at a fast pace, so that a specific product can be obsolete before the investment for a dedicated plant is recovered. This set of circumstances leads to the *multipurpose (MP) plant* as the major, basic configuration for fine chemical production. An MP plant has to be capable of handling a series of unit operations and performing many types of chemical reactions. The situation is different for active pharmaceutical ingredients (APIs)-for-generics.

Fine Chemicals: The Industry and the Business, Second Edition, by Peter Pollak
Copyright © 2011 John Wiley & Sons, Inc.

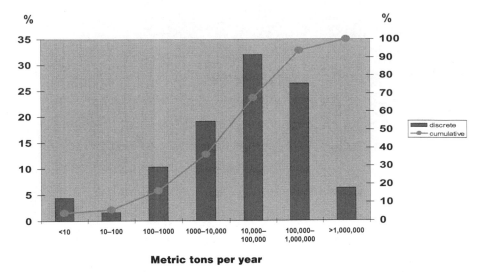

Figure 5.1 API production volumes (top 500 drugs).
Source: IMS Health.

As the markets are more predictable and stable, and production volumes are higher, dedicated plants are used often.

The main pieces of equipment used are agitated, jacketed vessels for carrying out the reaction, filters, or centrifuges for the solid/liquid separation and dryers. In the same plant, up to 20 or more different process steps can be executed per year.

Fine chemicals that have to be manufactured according to current Good Manufacturing Practice (cGMP) regulations include advanced ("regulated") intermediates and APIs for pharmaceuticals, which are manufactured via chemical synthesis, biotechnology, extraction, recovery from natural sources, or any combination thereof, as well as products such as veterinary drugs and vitamins. Separate regulations apply for food and feed additives, personal care products, and flavors and their advanced intermediates. The most demanding measures have to be taken for biopharmaceuticals, high-potency active ingredients (HPAIs), toxic substances, and drugs regulated under the Controlled Substances Act (CSA) by the U.S. Drug Enforcement Agency to stop illicit manufacture of narcotics, anabolic steroids, and similar compounds.

A helpful overview of different MP plant concepts can be found in Rauch [6].

Other plants, apart from the standard multipurpose, batch wise operated type, are used in the following cases:

- Single purpose (dedicated), batch wise operated plants are often used by the life science industries for large volume active ingredients. This is a.o. the case for many of the top ten pharmaceutical and agrochemical companies.
- Continuous plants are used for large volume fine chemicals made from gaseous or liquid raw materials.

- High potency active pharmaceutical ingredients (HPAPIs), mainly cyto-toxic PFCs with an occupational exposure limit below $10\,\mu g$ per m^3, are produced with bench scale equipment in glove boxes (see photo 13 in the insert).
- Microreactors are more and more used if one or more of the following conditions apply: strongly exothermic, endothermic, or otherwise hazard-ous reaction conditions, low volume requirements, gaseous or liquid reagents (see Section 4.1).

Before embarking on a project for a new plant, the following options for creating additional production capacity within existing plants should be con-sidered. They are listed in order of increasing associated cost:

- Outsourcing of part of the chemical manufacturing.
- Creation of additional capacity by process intensification, that is, increas-ing the efficiency of processes for existing products. On the technical side, this can be achieved either by developing processes that involve fewer steps, by increasing the yield, product concentration or reducing the reac-tion time, or by eliminating production bottlenecks, for instance, by install-ing external heating/cooling loops or separation units. Obviously, a combination of two or more of these measures is particularly effective. Options on the business side are subcontracting earlier parts of a multi-step synthesis to third parties, or "bottom slicing," namely eliminating the least profitable items from the product portfolio.
- Expansion of an existing plant by adding production bays, preferably in spare space in existing buildings. Thus, one takes advantage of existing infrastructure.
- A third possibility, which cannot be ranked in terms of financial attractive-ness, is the purchase of an existing fine chemical plant. Depending on the prevailing market conditions, the purchase price can be anywhere between a nominal fee and a large EBITDA multiple. Apart from the status of the fixed assets, a decisive element in the valuation of candidates is the good-will, particularly the existing manufacturing agreements and the solidness of the business plan. The lack of multipurpose capabilities is often an issue.

The *feasibility study* represents the first step in a design phase. A task force, consisting of process engineers, sales & marketing representatives, and other specialists, led by a project champion, develops the definition of the project and a first cost estimate. Typically, the project champion will be responsible for implementing the project. Already in this very first design phase appropri-ate measures have to be taken, if the plant needs to operate according to cGMP rules. Already the design itself has to undergo a qualification process, namely, the design qualification (DQ). The qualification process is a procedure proving and documenting that equipment and ancillary systems are properly installed, works in accordance to the design definition, and actually lead to the expected results. The overall qualification process consists in general of the following steps:

- *User requirement specification* (URS)—documented definition of the project.
- *DQ*—documented verification that the proposed design of the system is suitable for the intended purpose.
- *Installation qualification* (IQ)—documented verification that the system, as installed or modified, complies with the approved design.
- *Operational qualification* (OQ)—documented verification that the system performs as intended throughout the anticipated operating ranges.
- *Performance qualification* (PQ)—documented verification that the system, as connected together, can perform effectively and reproducibly according to the approved process method and specifications.

After alternatives have been checked and the definition of the project is found to be acceptable, the next design phase, the *basic design*, is initiated. At this point of time, it is appropriate to involve an external contractor. Engineering companies that are experienced in designing and building fine chemical plants are, a.o., Chemgineering, Switzerland; Chemieanlagenbau Chemnitz, Germany; Foster-Wheeler, USA; InfraServ Knapsack, Germany; Jacobs Engineering, USA; NNE Pharmaplant, Denmark.

The result of the basic design phase is a rather precise plan of the project and an accurate cost estimate that will constitute the basis for the final go/ no-go decision. The environmental impact of the project and all relevant permitting issues need also to be resolved during this phase. The detail engineering finally will provide the necessary information needed to execute the project.

In the design of a fine chemical plant, the *number of components* and *size of the equipment*, especially the volume of the reaction vessels, are critical. In order to ensure that the potential customer's needs are met by the capabilities of the plant, this has to be defined in close coordination with the marketing and sales (M&S) function. Depending primarily on the differing quantities of the fine chemicals to be produced in the same multipurpose unit, the concentration of substances in the reaction mixture, and the reaction time, there is, however, an upper limit for the size of the reaction vessel and the ancillary equipment. Some factors run countercurrently to the economy of scale and point to small-size equipment:

- Length of the production campaign—if it becomes shorter than about 10 working days, the changeover time for preparing the plant for the production of the next product becomes too long and overburdens the production costs.
- Working capital—if the equipment is oversized with regard to the requirement for any particular fine chemical, the interval between two production campaigns becomes too long, and excessive inventory is built up.

- Heat transfer—the time required for heating and cooling the reaction mixture and for its transfer among different pieces of equipment becomes too long as compared to the reaction time as such.
- In the case of expensive fine chemicals, the value of one batch in one piece of equipment becomes very high, sometimes in excess of $1 million, and therefore the risk of false manipulations becomes excessive.
- The dimensions of existing buildings, tank farms, and the capacity of utilities often determine an upper limit of the equipment size.

In commercial plants, the volume of the reactors ranges typically between 4 and $6\,m^3$ (sometimes between $1\,m^3$ and $10\,m^3$, or in rare cases, even larger). As a rule of thumb, the annual capacity for a one-step synthesis process averages approximately 15–30 MT of product per $1\,m^3$ reactor volume. Therefore, a production bay, which is equipped with 4 and $6\,m^3$ reaction vessels, is suitable for the production of around 100 MT of a step per year. As illustrated in Figure 5.1, this corresponds to a typical production volume of an API. Whereas one-third of the top 500 drugs are produced in the volume range of 10–100 tons per year, the requirement of 7% of the APIs exceeds 1000 tons per year.

Standard reaction conditions and standard *materials of construction* available in MP plants are usually:

Temperature	−20°C up to 200°C
Pressure	10 mbar–3 bar
Material of construction	Stainless steel and glass-lined

These conditions allow for the vast majority of production processes (see Section 4.1). In order to make an MP plant really fit for today's broad market requirements, an extension of the standard conditions by adding special features to enhance the flexibility of a plant is an absolute must. Flexibility, however, always has its price. Exotic or highly specialized equipment should to be installed only in an MP plant, if there is a specific need. Excessive flexibility is counterproductive. In the industrial practice, it has proved to be a good solution to provide space for special equipment in the basic design, and to order and install it only in case of a real demand.

Examples of special equipment for the *wet section* of the plant are low-temperature or cryogenic reactors, allowing for temperatures as low as −100°C, high-temperature reactors (up to +300°C), and high-pressure reactors, allowing operation at pressures up to 100 bar, fractional rectification columns, thin-film evaporators, liquid–liquid extractors, and various types of chromatographic columns (solid-phase, simulated moving bed [bi], supercritical fluid [SCF]). Beside traditional stainless steel and glass lining as materials of construction, more exotic materials such as hastelloy, tantalum, zirconium, and inconel alloys are increasingly used.

In the *dry section* of the building, micronization equipment, conventional dryers, nutsche–dryer combinations, spray dryers, air classifiers, packaging/labeling machines, and other equipment can be considered.

Instead of adding special equipment to individual production bays, it is also possible to place them centrally in the dry or wet section of the plant, respectively. In this way, they can be connected as needed to different production bays. Another option is to create semi-specific production bays, For example, for hydrogenations, phosgenizations, Friedel–Crafts alkylations, and Grignard reactions.

The choice of the proper *piping* concept is essential for a valid MP plant design. The basic requirements for a piping system are, beside corrosion resistance for a wide array of substances, ease of cleanability (due to quality and costs) and, of course, a high degree of flexibility in order to ensure the needed multipurpose character of the plant. Typically, the following approaches are available:

· A preinstalled piping system with an adequate number of manifolds and coupling stations, according to the required flexibility (see Fig. 5.2). This classical system may be advantageous in cases where the product mix tends to be not too broad and/or the number of product changes per unit of time is relatively small.

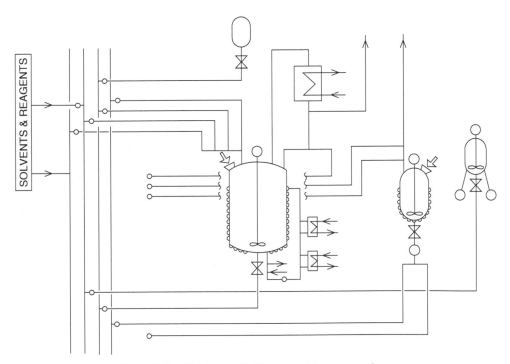

Figure 5.2   Piping manifolds for multipurpose plants.

- A process-specific piping concept is the system of choice in cases where the products to be manufactured are still unknown during the design phase of the plant. This system is also ideal in cases when the campaign lengths are expected to be short, that is, when frequent product changes are likely. The process-specific piping concept generally minimizes the needed amount of fixed-installation pipes. Connections between reactors, head tanks, receivers, pumps, filtration units, and other components are installed only as needed, on a strictly campaign-to-campaign basis. In addition, suitable hoses are installed instead of solid piping whenever possible. This concept also facilitates the cleaning–changeover process, as it minimizes or even avoids "deadlocks."

The complexity of the plant design, the degree of sophistication, and the quality requirements of the fine chemicals to be produced; the necessity to process hazardous chemicals, the sensitivity of product specifications to changes of reaction parameters, and the availability of a skilled workforce all determine the degree of *automation* that is advisable. Full process control computerization for an MP plant is much more complex than for a dedicated single-product plant and therefore will be also much more expensive. Whenever possible, all efforts have to be made to choose standard process control systems and to apply standard control software; this is a proven measure to control the investment costs in this segment and will also minimize the risk of having excessive investment and start-up costs due to initiating problems with the computer control system.

The fact that automation systems need to be validated has become a critical aspect of all automation systems that are being applied for cGMP productions. Some guidelines on this topic can be found in the U.S. *Code of Federal Regulations* [7a]. Good manufacturing guidelines apply also to heating, ventilation, and air-conditioning (HVAC) systems. They have to be designed to exclude contamination.

The *material handling* in an MP plant is driven mainly by the following considerations:

- To optimize direct labor costs versus investment costs by the mechanization of material handling operations.
- To comply with all pertinent quality requirements regarding safety, hygiene, and cGMP, if applicable.

According to the nature of the substances involved, specific segregation within the production area might be necessary. In order to exclude the risk of any cross-contamination, the following precautions might be taken:

- Dispensing of starting materials and charging of solids into reactors might be located in isolated and contained areas.

- The transfer of wet solid material from centrifuges or nutsches to dryers should occur either via closed transfer pipes or via a solid material tote bin system.
- Depending on the nature of the products, the unloading of dryers might have to take place in a segregated area (e.g., clean rooms for cGMP products).

In case of cGMP productions, both the people and material flow have to follow strict rules. Specifically, the following activities have to be fully integrated into the material flow process:

- Receipt, identification, sampling, and quarantine of incoming materials, pending release or rejection
- In-process control laboratory operations
- Sampling and quarantine before release or rejection of intermediates and APIs
- Holding rejected materials before further disposition (e.g., return, reprocessing, or destruction)
- Storage of released materials
- Packaging and labeling operations

Even during the very early design phase, people and material flow has to be modeled in order to identify these requirements.

Three examples of state-of-the-art MP plants are described in Figures 5.3–5.5. They represent (1) a typical pharmaceutical fine chemical (PFC) plant of a small Swiss company, Rohner-Chem, Switzerland; (2) a large fine chemical plant with an innovative layout built by the German pharmaceutical company Schering AG (now Bayer Schering Pharma) for its captive API requirements (see photo 2 in the insert); and (3) the equipment scheme of a production bay from a concept study, "Fine Chemicals Complex 2" (FCC-2) in Lonza's Visp, Switzerland plant.

Operating principles of MP plant example 1 (Fig. 5.3) are as follows:

- *Bay concept*—the logical operating unit of the plant is a bay. A typical bay consists of two to three multipurpose reactors (up to a maximum volume of $10 \, m^3$ each), one filtration unit (nutsche or centrifuge), and one dryer.
- *Production flow of the plant*—charging of starting materials (level 4), reaction (level 3), crystallization (level 2), filtration (level 1), and drying and blending (level 0).
- *Material flow* area—reserved zone for material flow.
- *Open structure*—manufacturing in a maximum flexibility and minimal segregation environment, six bays in the same area. The reactors and

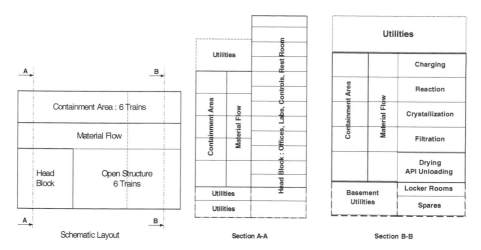

Figure 5.3 Multipurpose plant example 1.
(*Source*: Rohner AG [Beteiligungsgesellschaft Arques] at Pratteln, Switzerland [15].)

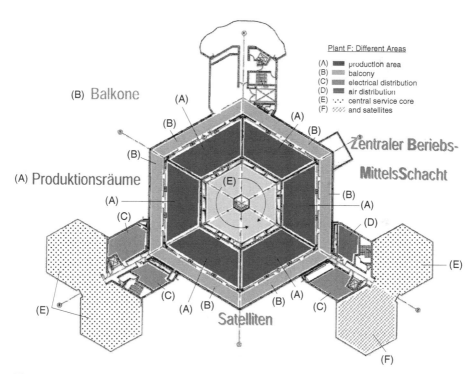

Figure 5.4 Multipurpose plant example 2.
(*Source*: Schering AG, Bergkamen, Germany.)

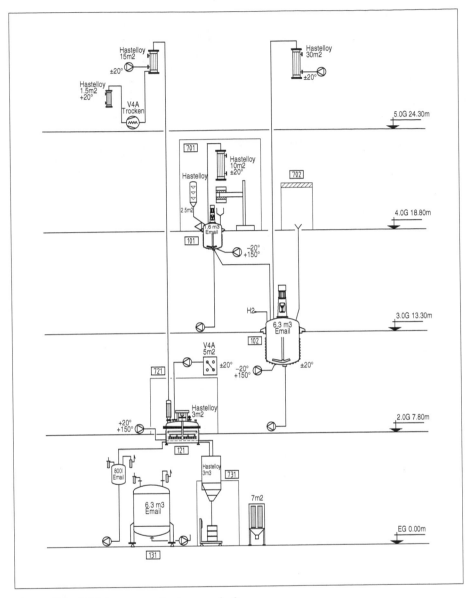

Figure 5.5 Multipurpose plant example 3.
(*Source*: Lonza, Switzerland.)

filtration units of the different bays can be connected as needed. This approach also allows for a maximum capacity utilization.

- *Containment area*—manufacturing combined with maximum segregation; six compartments, each housing one bay.

- *Headblock*—containing in-process control laboratories, offices, locker, and meeting rooms, integrated into the main structure of the building.
- *Infrastructure*—chiller, off-gas treatment, air-conditioning systems, process water, spares, and other facilities, located in the basement or as open-air installations on the roof of the plant.
- *Manufacturing standard*—cGMP, intermediates, key intermediates, and APIs.

Operating principles of MP plant example 2 (Fig. 5.4) are as follows:

- The futuristic-looking hexagon-shaped plant design with satellite buildings is the result of a new developed safety–ecology–operating concept.
- The building complex, which tops 42 m in height, has a diameter of 88 m, and a working area of over 28,000 m^2, is operated by over 100 well-bayed chemical operators, engineers, and chemists.
- The satellite buildings contain service areas, laboratories, storage areas, offices, and various utilities (ventilation, electricity, brine, steam, inert gas, and water for fire protection).
- For the processing flow, a top-down approach was chosen, utilizing gravitational force whenever possible.
- The plant houses six segregated and independent manufacturing areas, in order to separate, for example, corrosive chemistry from final purification steps of APIs.
- Production takes place in strictly closed equipment and is controlled by a state-of-the-art process control system.
- The core of the hexagon-shaped building is used for the central services, and supply of liquid and gaseous media via a ring pipe.
- Manufacturing standard: cGMP, intermediates, key intermediates, and APIs.

Operating principles of MP plant example 3 (Fig. 5.5) are as follows:

- The building is subdivided vertically in 10 bays for PFC/API production and horizontally into five floors. On floor 5, a condenser and a vacuum pump for the evacuation of the "nutsche" are installed, as well as an overhead condenser for the reaction vessel (see text below). Floor 4 houses a contained cabin for charging solid raw materials and intermediates and an agitated 1.6-m^3 feeding tank for catalyst slurries. The latter is connected to a plate filter for catalyst and charcoal recovery, a filling station for big bags, and an overhead condenser. The jacketed 6.3-m^3 glass-lined reaction vessel is installed on floor 3. The main piece of equipment is a hastelloy filter nutsche (Ø 2100 mm) for liquid–solid separation. The mother liquor is discharged to a holding tank. The unit is connected to a heating/cooling temperature control module (TCM). The filter cake is discharged to a silo and then filled in big bags or drums. It is further processed in a separate drying–sieving–filling station.

- The building will also contain absorption columns for the pretreatment of waste gases as well as distillation columns for solvent recovery. A generous infrastructure is available on site. Next to the building, a tank farm for solvents with 25- and 50-m^3 tanks will be built. Storage of intermediates and finished products takes place in a state-of-the-art warehouse. For fire protection, the air in the warehouse is diluted with nitrogen, thus reducing the oxygen concentration to 13.2%. Utilities comprise electric power from a nearby hydroelectric power station, steam generated in the waste incinerator, and nitrogen from an onsite air separation plant.

The production of fine chemicals using *biotechnological processes* fundamentally follows the same pattern as the one for synthetic fine chemicals: preparation and charging of the raw material, reaction, liquid–solid (crude product) separation, product purification, and packaging. Depending on the specific bioprocess used, there are, however, generally substantial differences in the design and operation of the plant. Simple fermentations used for specific steps in low-molecular-weight fine chemicals (e.g., conversion of a carbonyl to an amido group, or of a carbonyl to a chiral hydroxy group) can be carried out in conventional MP plants. This is particularly the case if enzymes fixed on a solid support are used as catalysts. The production of modern high-molecular-weight biopharmaceuticals by the use of recombinant processes requires specifically designed plants, where utmost attention is paid to the safeguard of sterility (see Section 4.2).

Fine chemical plants have only evolved in few aspects and discrete steps such as containment and automatic process control over the past 25 years. Different initiatives for radical improvements are underway. With modular MP plants, an efficient combination of the flexibility of batch plants with the high performance and easier scale-up of continuous flow chemistry is sought. The "European Roadmap for Process Intensification" [8] has formulated efficiency targets with timescale and performance estimation. The study has evaluated 72 process intensification technologies; 46 of them have been retained for a detailed evaluation and description in technology reports. Hessel et al. review application examples [9]. Along the same lines, the FDA has started an initiative toward knowledge-based processing called "Quality by Design" It is summarized in the report "Pharmaceutical Quality for the 21st Century: A Risk-Based Approach" [10].

Another visionary project is the "F³ Factory" [11]. It means Flexible, Fast, and Future Factory and is an ambitious private–public project aimed at developing the chemical plant of the future, which is capable of widespread implementation throughout the chemical industry. The plant incorporates advances in process intensification concepts and modular "plug-and-play" chemical production technology. It is targeted to be more economic, eco-efficient, and sustainable than conventional processes, both in continuously operating large-scale plants and in small- and medium-sized batch plants. In order to demonstrate the technical feasibility, a modular continuous plant is being built at Chempark Leverkusen, Germany. The leading Consortium has a budget of €30 million (of which €18 million is funded by the European Union), a project duration of 4 years (2009–2013). It consists of a dozen major chemical and pharmaceutical companies (a.o. Arkema, AstraZeneca, BASF, Bayer TS, Evonik-Degussa, Procter & Gamble, Thodia) and 15 universities (a.o. Centre Nationale de Recherche Scientifique, CNRS, France; KTH Royal Institute of Technology, Sweden; University of Newcastle, UK; Ruhr-Universität, Germany; Technische Universiteit Eindhoven, The Netherlands).

5.2 PLANT OPERATION

Optimum capacity utilization of capital-intensive MP plants is crucial for the profitability of a fine chemical company. Typically, an investment of more than $1 is required to generate $1 of sales. The actual capacity utilization in a given time period is determined both by business and technical factors. The former relate to the size and quality of the order backlog, which has to be taken care of by M&S, the latter to the capability of making the best use of the equipment, which is the responsibility of plant operation). Conflicting interests of M&S, production, procurement, and controlling have to be aligned carefully. Particularly critical is an excellent communication between M&S, which determines what quantity of which products can be sold; and production, which determines how the existing equipment can be used most advantageously and what type of plant is needed in the future. There are both short- and long-term aspects to production planning. A useful tool for short-term planning is a rolling 18-month sales forecast, with binding commitments for the first 2–6 months and more flexibility for the rest of the periods.

The traditional notion of capacity utilization was based on a one-dimensional "yes/no" approach, assuming full utilization, when the plant is running, and zero utilization, when it is not. Taking advantage of the experience of the automotive industry, which is the forerunner in *lean production*, a much more sophisticated methodology, called *Overall Equipment Effectiveness* (OEE) [7], is gaining ground also in the fine chemical industry. OEE considers the expenditure of resources for any goal other than the creation of value for the customer to be wasteful, and thus a target for elimination. As different products with widely varying throughputs in terms of kilograms per annum are produced during the course of a year, production planning is a very demanding task. To begin with, it is not possible to indicate a production capacity for an MP plant in terms of "kilograms, or tons, of product per year," as is the case in commodity chemical industry, which runs dedicated (mono product) plants. As a kind of an artifact, the criterion "m^3 of reactor volume" has been chosen as a reference unit instead. It is, however, a rather imprecise measure as neither capacity per se nor capacity utilization are defined unambiguously. With regard to the former, already the type of equipment, which determines the reactor volume, is not well defined: Does one consider only the reaction vessels as such or also crystallization vessels and buffer tanks? Furthermore, in MP plants, some pieces of equipment may not be used at all for the production of a specific fine chemical. This is for instance the case, when a liquid product is produced in a bay that is equipped with an (expensive) filter-nutsche. OEE allows determining how a production bay actually performs in comparison to an ideal one, which runs at maximum throughput, without production interruptions, and without reworks of out-of-spec product. As shown in Figure 5.6, OEE is determined by three factors, namely (plant) availability (i.e., the fraction of time the plant is actually running), performance (i.e., actual versus theoretical throughput), and quality (i.e., released product as compared with total product made). In practice, OEE values are much lower than expected. In the figure it is only about 33%. The arduous task of improvement of OEE begins with a meticulous search, and documentation of commercially available software is becoming increasingly accessible, which efficiently supports the complex task of production planning in MP plants. Plant operators play an important role in reducing lost time, which can be caused for instance by late arrival of starting material, instable processes, long changeover times, delays in in-process testing, mechanical deficiencies of equipment (e.g., agitators and pumps), or a temperature alarm.

In today's global economy, it is vital for fine chemical manufacturers to adhere to *international standards for safety and ecology*. For that purpose,

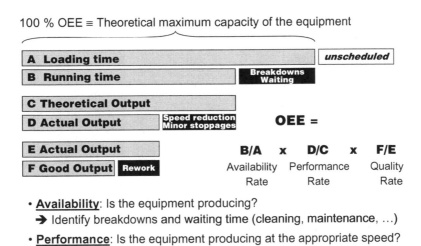

Figure 5.6 Overall equipment effectiveness.

there are several highly developed systems available, including the International Organization for Standardization's *ISO management system*, ISO, Geneva; the *Responsible Care trademark of the American Chemistry Council* program, which is of U.S. origin; or the European Union *Eco-Management and Audit Scheme* (EMAS), European Commission, Environment DG. The *ISO 14001* set of *ISO's management system* standards focuses on minimizing harmful effects on the environment and achieving continuous improvement of environmental performance. *Responsible Care* is a voluntary program, initiated be the U.S. chemical industry, to achieve improvements in environmental, health, and safety performance beyond levels required by the U.S. government. *Responsible Care* continues to strengthen its commitments and enhances the public credibility of the industry. Finally, the *Responsible Care 14001* certification process combines *ISO 14001* with the *Responsible Care* program; for instance, the revised EMAS includes the *ISO 14001* system.

In order to safeguard human health and the environment, REACH (Registration, Evaluation, Authorization and restriction Chemicals) regulation entered into force in Europe on June 1, 2007. Under REACH provisions, all producers and importers of chemicals in the European Union have to register their products with the European Agency for Chemical Substances (ECHA) in Helsinki. About 30,000 substances existing prior to 1981 and 5000 new substances produced or traded in Europe in quantities of more than 1 ton per year will be affected. The deadline for the submissions of data for chemicals produced or traded in the range of 1–100 tons/year is the end of May 2018. PFCs produced and processed under strictly controlled conditions are exempt. In order to reduce animal testing, manufacturers of the same chemicals must pool their animal tests.

Executives list REACH as the most burdensome regulation of all. They reckon that with costs between €2.8–€5.2 billion during the 11 years that introductory extensive tests are required for the authorization, the financial burden for the chemical industry will be substantial (for details see Section 14.1, Western Hemisphere). U.S. Congress eyes its own REACH legislation. Most likely efforts will focus to revise the Toxic Substances Control Act (TSCA), enacted in 1976.

Quality and documentation aspects in general have become an increasingly important success factor in the fine chemical business. This is even truer for cGMP production. Because fine chemicals are sold according to stringent specifications, adherence to constant and strict specifications, at risk because of the batchwise production and the use of the same equipment for different products in MP plants, is a necessity for fine chemical companies. The *ISO management system standards*, which are implemented and recognized world-wide, play an important role. Specifically, the *ISO 9001* family deals primarily with quality management and focuses on the customer's quality requirements, regulatory requirements, customer satisfaction, and continuous improvement on all pertinent processes.

Fine chemicals intended for use in pharmaceuticals are to be manufactured according to guidelines for industry *ICH Q8* [12], which is GMP for active pharmaceutical ingredients. These guidelines were developed within the Expert Working Group of the International Conference on Harmonization (ICH) of the technical requirements for registration of pharmaceuticals for human use. Since 2001, the document has been applied by the regulatory bodies of the European Union, Japan, and the United States. In addition, the United States *Code of Federal Regulations* [13] represents guidelines specific for the United States. A firm producing pharmaceuticals has to be approved by national authorities. If the products are intended for the U.S. market, an inspection of the premises and the relevant documentation by the FDA is also required. The inspectors use three classifications for their observations, namely NAI (no action indicated), if the firm is compliant, VAI (voluntary action indicated), if a firm has several violations that have to be corrected as soon as possible, and OAI (official action indicated), if the findings are significant. In the latter case, they usually are documented in the "warning letter" and require immediate attention to prevent an injunction, seizure, and/or prosecution.

General standards for drugs are typically published in the so-called national pharmacopoeia. The names of the different national pharmacopoeia are formed by pharmacop(o)eia combined with the name of the country, for example, the *United States Pharmacopeia and National Formulary* (USP–NF). Attempts to generalize and unify the different national pharmacopoeia have continued for over a century. The European Community signed a convention that resulted in issuance of the *European Pharmacopoeia* [14]. Finally, the United Nations World Health Organization (WHO) publishes a *Pharmacopoeia Internationalis* [15].

Standards for food-grade chemicals in the United States are published in the *Food Chemicals Codex* (FCC) [16], for laboratory reagents in *Reagent Chemicals—ACS Specifications* [17], and for electronics-grade chemicals in the *Book of SEMI Standards* (BOSS) by Semiconductor Equipment and Materials International (SEMI). The latter two product categories, with the exception of reagent chemicals used as diagnostics, are not subject to cGMP regulations.

A comprehensive training program for all employees is another essential element to secure adequate quality and safety standards. The program has to incorporate the entire workforce involved into any aspect of the manufacturing process and needs to be documented.

All quality aspects within a company are to be controlled by an independent organizational unit. Beside the quality control unit, of course, the quality assurance activities are also part of this operation. The following main aspects are considered here:

- Releasing or rejecting products.
- Reviewing and approving qualification reports.
- Reviewing and approving validation reports. The validation process is a program that provides a high degree of assurance that a process will

consistently produce a result meeting predetermined acceptance criteria.

- Approving all specifications and master production instructions.
- Making sure that critical deviations are investigated and resolved.
- Establishing a system to release or reject raw materials and labeling materials.
- Approving changes that potentially affect intermediate or API quality.
- Making sure that internal quality audits are performed.
- Making sure that effective systems are used for maintaining and calibrating critical equipment.

These criteria are mandatory for cGMP products; however, it is recommended practice to utilize, whenever possible, the same criteria for non-cGMP products as well.

A new program at FDA called *process analytical technology* (PAT) allows the use of continuous process control systems that measure and assess quality during the manufacturing process rather than between batches [18]. The framework specifies the development of manufacturing processes that can consistently ensure a predefined quality at the end of the manufacturing run. Such procedures would be consistent with the basic tenet for "quality by design" inherent in currently available commercial control systems.

CHAPTER 6

Research and Development

The overall emphasis of fine chemical R&D is more on development than on research ("small r, big D"). The main tasks are (1) designing, respectively duplicating and adapting in case of contract manufacture, and developing laboratory procedures for new products or processes; (2) transferring the processes from the laboratory via pilot plant to the industrial scale (the scale-up factor from a 10-g sample to a 1-ton batch is 100,000); and (3) to optimize existing processes. At all times during this course of action, it has to be ensured that the four critical constraints, namely, economics, timing, safety, ecology, and sustainability are observed.

The fine chemical industry is R&D intensive; expenditures are higher than in the commodities industry. They represent around 5–10% versus 2–5% of sales. On the business side, product innovation must proceed at a more rapid pace, because life cycles of fine chemicals are shorter than those of commodities. Therefore, there is an ongoing need for substitution of obsolete products. The growth of the business as such can kick off only once this backlog is filled. On the technical side, the higher complexity of the products and the more stringent regulatory requirements absorb more resources.

The *project portfolio* enables an overview on the ongoing research activities. Numerous economic and technical parameters have been proposed to provide a meaningful picture. Examples are attractiveness, strategic fit, innovation, gross/net present value, expected profits, R&D expenditures, development stage, probability of success, technology fit, potential conflicts with other activities of the company, and realization time. Most of these parameters cannot be determined quantitatively, at least during the early phases of a project.

The probability of success, for instance, depends on a number of factors. On the technical side, it is the likeliness that the laboratory results in terms of yield, throughput, and quality can be matched on the industrial scale within the planned timeframe. On the business side, it is the likeliness of realizing the forecasted sales and profit figures. In custom manufacturing, a twofold risk is incurred. On the one hand, the fine chemical company has to be chosen as

Fine Chemicals: The Industry and the Business, Second Edition, by Peter Pollak
Copyright © 2011 John Wiley & Sons, Inc.

a supplier; on the other hand, the customer, typically the pharmaceutical company, has to be successful in launching the new drug.

The best way to take advantage of a project portfolio is to develop and use it in an iterative way. By comparing the entries at regular intervals, for instance, every 3 months, the directions that the projects take can be visualized. If a negative trend persists with one particular project, the project should be put on the watch list.

6.1 OBJECTIVES

R&D has to manage the following functions in order to deliver the requested services.

Literature and Patent Research

An efficient literature and patent search capacity has to be made available. Patent research is particularly important for evaluation of the feasibility of taking up R&D for new active pharmaceutical ingredients (APIs)-for-generics. Provisions have to be made for a periodic examination of all acquired research results to safeguard Intellectual Property Rights (IPR) and to determine whether patent applications are indicated.

Process Research

This key function has to design new synthetic routes and sequences and undertake initial experiments to secure the feasibility. Two approaches are feasible. For simple molecules, the "bottom-up" approach is the method of choice. The researcher converts a commercially available starting material and sequentially adds more reagents until the target molecule is synthesized. The starting material is mostly a petrochemical commodity, sometimes a natural product. For more complex molecules, a "top-down" approach, also known as *retrosynthesis*, is chosen. Through the chemist's imagination, key fragments of the target molecule are first identified, then synthesized individually, and finally combined to form the desired molecule through convergent synthesis.

Process Development

This provision focuses on the design of new, efficient, stable, safe, and scalable synthetic routes to a target fine chemical. It represents an essential link between process research and commercial production. The resulting "base process" description provides the necessary data for the determination of preliminary raw material and product specifications, the manufacture of semi-commercial quantities in the pilot plant, the assessment of the ecological impact, the regulatory submissions and technology transfer to manufacture at industrial scale, and an estimate of the manufacturing costs in an industrial-scale plant. If the base process is provided by the customer as part of the technology transfer, process research has to optimize it so that it can be transferred to the bench-scale laboratory or pilot plant (see photos 5, 7, 16, and 18 in the insert). Furthermore, it has to be adapted to the specific characteristics of available production trains.

Bench-Scale Laboratory, Kilogram Laboratory, and Pilot Plant Development

Depending on the volume requirements, three different types of equipment are used for process research, development, and optimization, namely, *bench-scale laboratories* for gram to 100 gram, *kilo-labs* for kilogram to 10 kg, and *pilot plants* for 100 kg to ton quantities. The scale-up addresses the change in process conditions that arise from the greater dimensions and different geometries of industrial-scale reactors as compared with laboratory equipment. Apart from reduced mass and heat transfer rates (resulting primarily in longer reaction times), flow regimes, phase separation rates, interfacial surface areas, flow patterns, and heterogeneity in process streams are also dimensionally sensitive variables and parameters. Particularities of laboratory processes which have to be eliminated include the use of large numbers of unit operations, dilute reaction mixtures, vast quantities of solvents for extraction, evaporation to dryness, and drying of solutions with hygroscopic salts. The execution of the tasks associated with scale-up requires the skills of a generalist, who is both knowledgeable about laboratory synthesis and industrial production processes. In Anglo-Saxon countries, where the curricula of chemists and chemical engineers are different, this occasionally causes communication problems. Although modern reaction calorimeters consent to foresee the effects of these different conditions to a certain extent, a direct transfer of a process from the laboratory to the industrial scale is not recommended, because of the inherent safety, environmental, and economic risks. In development, the viability of the process on a semicommercial scale has to be demonstrated. Trial quantities of the new fine chemical have to be manufactured for market development, clinical tests, and other requirements. The necessary data have to be generated to enable the engineering department to plan the modifications of the industrial-scale plant and in order to calculate production costs for the expected large-volume requirements. Both equipment and plant layout of the pilot plant reflect those of an industrial multipurpose plant, except for the size of reaction vessels (bench-scale laboratory ~10–60 L; pilot plant ~100–2500 L) and the degree of process automation. Simplicity, robustness, short cycle times, and low costs require particular attention. Before the process is ready for transfer to the industrial-scale plant, the following activities have to be completed: adaptation of the laboratory process to the constraints of a pilot plant, (*haz*ard and *op*erability) HAZOP analysis, and execution of demonstration batches. The

TABLE 6.1   Laboratory Synthesis versus Industrial-Scale Process

Task	Laboratory Synthesis	Industrial-Scale Process
Operator	Laboratory chemist	Chemical engineer
Economy	Yield	Throughput ($kg/m^3/hour$)
Units	G, mL, mol, min. h	Kg, ton, h, shift
Equipment	Glass flask	Stainless steel, glass linedl
Process control	Manual	Automatic [reaction vessel]
Critical path	Reaction time	Heating/cooling
Liquid handling	Pouring	Pumping
Liquid/solid sep.	Filtration	Centrifugation

main differences between laboratory synthesis and industrial-scale production are shown in Table 6.1.

In case of current Good Manufacturing Practice (cGMP) fine chemicals, also a process validation is required. It consists of the three elements of process design, process qualification, and continued process verification.

Process Optimization

The involvement of R&D does not end once a new chemical process has been introduced successfully on an industrial scale. Process optimization is now called upon to improve the economics. As a rule of thumb, it should be attempted to reduce the cost of goods sold (COGS) by 10–20% every time the yearly production quantity has doubled. The task extends from fine-tuning the currently used synthetic method all the way to the search for an entirely different second-generation process. The guiding principle is improving the Overall Equipment Effectiveness (see Section 5.2). Specific provisions are

- Reduction of the number of chemical steps, a.o. by avoiding isolation of intermediates or substituting chemical with enzymatic steps (see Section 4.2)
- Reduction of raw material costs by yield maximization or substitution with cheaper alternatives
- Reduction of solvent consumption by higher concentration and/or recycling.
- Reduction of catalyst and biocatalyst usage, a.o. by recycling
- Reduction of the environmental impact, that is, making the process "greener"

An important tool for assessing the environmental impact of a process is the *atom economy*, that is, the ratio between the atoms contained in the desired product and the total number of atoms introduced into a process (raw materials and solvents). The calculation of atom economy is quite demanding. For multistep processes, Environmental Assessment Tool for Organic Synthesis (EATOS) enables trade-offs of yield versus environmental impact to be readily assessed.

GlaxoSmithKline has a 2015 waste reduction target for its plants of 30 kg for each 1 kg of API, down from 100 kg in 2005.

Ancillary tasks are

- *Analytical Development.* The increasingly complex molecules require a permanent development of new, sophisticated, analytical methods and, if required, their validation. In order to fulfill this demanding task, a well-equipped state-of-the-art analytical laboratory has to be accessible. In custom manufacturing projects, a close cooperation between the analytical departments of the customer and the supplier is essential for a successful completion of a project.
- *Hazard Analysis and Risk Assessment.* Before a new or revised process is cleared for pilot and—later on—industrial-scale production, it has to pass a detailed, standardized HAZOP analysis. HAZOP comprises detailed safety reviews of the chemical and mechanical processes involved, such as reactions, distillations, rectification, drying, and milling–blending operations.

Fine chemicals made by fermentation or natural product extraction processes are not burdened by a broad range of route possibilities. Bench development by microbiologists and engineers proceeds along the following steps [19]:

- Microbial or plant cell fermentations:
  1. Elucidation of the pathway to the secondary metabolite
  2. Nutrient, precursors, and optimization of fermentation cycle conditions
  3. Strain and cell line improvements with respect to productivity and robustness in fermentation
  4. Data gathering to support scale-up to stirred tanks at different scales
  5. Definition of the downstream process candidate for recovery, concentration, purification, and isolation of the target molecule from fermentation
- Extraction from natural sources (plant or animal material):
  1. Evaluation of differing sources of the compound-bearing materials
  2. Pretreatment conditions for successful extraction
  3. Extraction or leaching conditions, solvent, or extracting stream material selection, and separation of spent plant material

4. Definition of the process candidate for concentration, purification, and isolation

5. Data gathering to support scale-up

Most likely, both technologies eventually have to deal with relatively large volumes of cell mass or plant material waste, and benchwork to address those issues is also needed.

6.2 PROJECT INITIATION

There are two main sources of new research undertakings: concepts originating with the researchers themselves ("supply push") and those coming from customers ("demand pull"). Ideas for new processes typically originate from researchers, ideas for new products from customers, respectively. Particularly in custom manufacturing, "demand pull," promoted by business development (see Section 12.1), prevails. Incoming ideas for new research programs have to be evaluated at regular intervals, and decisions have to be taken as to whether to pursue or reject an idea and initiate an R&D project. The main criteria for the selection of *new product ideas* are the size and attractiveness of the demand on the one hand, and the fit with the supplier's unique selling proposition (USP) on the other hand. What is expressed bluntly with the saying "Can we sell it, can we make it, will we make money?" translates into a meticulous pondering of business, technical, and financial considerations in the industrial reality.

The "new product committee" is the body of choice for this task. It has the assignment to evaluate all new product ideas. It decides whether a new product idea should be taken up in research, namely, whether it becomes a project (see discussion below).

The tasks and members are described in Section 12.1. A very comprehensive "project assessment checklist" has been published [20]; a simplified version is reproduced in Appendix A.2. It is not recommended to summarize the attractiveness of a project in a single number—and to reject project B, which totals 3.9 versus project A, which totals 3.8 points. It could just be that B had a higher-ranking in patentability, but a lower ranking in competitiveness of pricing than A! Last but not least, the committee decides also about the abandonment of a project, once it becomes evident that the objectives cannot be reached.

6.3 PROJECT EXECUTION AND MANAGEMENT

Once a new idea has been accepted, a *project management* organization is set up. Its task is to successfully complete the project by defining the objectives that have to be met, the human and material resources needed, and the mile-

TABLE 6.2   Duties of the Project Champion and Project Manager

	Project Champion	Project Manager
Key responsibility	Successful completion of project overall	Successful completion of technical aspects
Subsidiary responsibilities	• Mgmt. of project team • Financials (budget, investments, etc.) • Customer relationship management (CDA, R&D agreement, supply contracts) • Pricing • Market and competitor intelligence	• Development of a viable, scalable, economic, and environmentally safe process • Technology transfer • Preparation of samples, trial quantities, and validation batches • Process validation • Preparation of regulatory submissions
Assigned functions	• Business development • Controlling • Legal • Procurement	• R&D • QA/QC • Engineering • SHE

stones to be reached, and by establishing the internal and external reporting system. A detailed project schedule for a custom manufacturing project is reproduced in Appendix A.3. The project team is cross-functional and comprises representatives from most business functions. They can be on either full- or part-time assignments. In a typical project, the overall responsibility for the economic and technical success lies with the project champion. If the project is executed for a key account, the pertinent key account manager is the prime candidate for the position. This individual is assisted by the project manager, who is responsible for the technical success. All technical project team members report to the latter. The duties of the two key persons are listed in more detail in Table 6.2. While individual team members may have customer contacts for specific questions, such as the preparation of the reference analytical method, the main information flow between customer and supplier goes through the project champion. The new project committee overlooks the project and coordinates with the R&D and business development departments (see also Section 12.1).

In custom manufacturing, a typical project starts with the acceptance of the product idea, which originates mainly from business development, by the new product committee, followed by the preparation of a laboratory process, and ends with the successful completion of demonstration runs on industrial scale and the signature of a multiyear supply contract, respectively. The input from the customer is contained in the "technology package." Its main constituents are (1) reaction scheme; (2) target of project and deliverables (product, quantity, required dates, specifications); (3) list of analytical methods; (4) process development opportunities (stepwise assessment); (5) list of required reports;

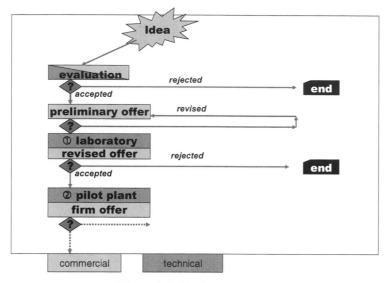

Figure 6.1   Project sequence.

(6) safety, health, and environment (SHE) issues; (7) materials to be supplied by customer; and (8) packaging and shipping information. The technical part of a project usually determines its duration. Depending on the quality of the information contained in the "technology package" received from the customer and the complexity of the project as such, particularly the number of steps that have to be performed; it can be any time between 12 and 24 months. Depending on the number of researches involved, the total budget easily amounts to several million U.S. dollars. Figure 6.1 gives an impression of the initial decision points that a project has to pass during its execution.

Key success factors for project management are

- Appointment of experienced managers with leadership qualities to the roles of project manager and champion. They also must be well viewed by the customer.
- A project plan with well-defined and clearly communicated roles, responsibilities, and timetable of deliverables must be agreed on.
- Project managers and champions must be empowered to manage scientists and other team members to execute the agreed-on project plan, to write reports, and so on.
- Customers must be encouraged to have direct contact with project managers.
- A provision for safeguarding confidentiality and intellectual property rights must be included.

- Modifications of the project plan must be kept to a minimum and be adequately communicated.
- Proper follow-up and interaction with customers on all project-related matters, including the preparation of progress reports according to an agreed-on format, should be ensured.
- Adequate software tools, such as Microsoft Project, Excel, Lotus Notes, and Primavera, should be employed.

■■■■■■ **CHAPTER 7**

Cost Calculation

7.1 INVESTMENT COST

Investment costs for multipurpose plants vary considerably, depending on the location, size of equipment, and degree of sophistication (e.g., automation, containment, quality of equipment). An example for a current Good Manufacturing Practice (cGMP) multipurpose plant built in the United States is shown in Table 7.1. The investment cost of $21 million comprises just the equipment and installation. The building, property, and external services are excluded. For comparison purposes, the investment cost per cubic meter reactor volume is used. In this case, it is $0.9 million. The amount includes the cost of the reaction vessel itself plus an equitable part of the ancillary equipment, such as feeding tanks, piping, pumps, and process control. Costs for a plant that is used for the production of nonregulated intermediates only would be substantially lower. Pharma companies tend to spend two to four times more for a plant with the same capacity. In contrast, investment costs in developing countries, particularly in India or China, would be considerably lower (see Chapter 14).

A breakdown of the investment costs according to major categories is shown in Table 7.2. The variation in the voice "piping" is determined, a.o. by the piping concept (preinstalled or process-specific; see Section 4.1) and the sophistication of the insulation; in "equipment," it depends on the complexity of the individual items. Costs of reactors, for example, vary according to the material of construction, the operational pressure and temperature ranges, the heating/cooling system (jacket or coils), and the construction of the agitating system. In contrast, the impact of the equipment size on the total investment costs is modest. In the example given in Table 7.1, the size of the reactor is $4\,m^3$. If larger or smaller reactors were installed, the unit cost per cubic meter would decrease or decrease with the exponent 0.5, respectively. Thus, an otherwise comparable plant equipped with six $2\,m^3$ reactors would require an investment of $15 million. Hence, by increasing the equipment size, manufacturing costs on a per

Fine Chemicals: The Industry and the Business, Second Edition, by Peter Pollak
Copyright © 2011 John Wiley & Sons, Inc.

TABLE 7.1   Investment Cost for a cGMP Multipurpose Plant

Equipment/Investment	Numbers
Description of main equipment	
Production trains	2
Reactor vessels (volume = $4\,m^3$)	6
.... Total reactor volume	.... $24\,m^3$
Filtration units	2
Dryers	2
Capital investment	
Total capital investment	$21 million
• Investment per production train	$11.5 million
• Investment per piece of main equipment	$ 2.1 million
• Investment per cubic meter reactor volume	$ 0.9 million

TABLE 7.2   Major Investment Cost Categories of a Multipurpose Plant

Investment Category	Share
Piping and installation (including insulation and painting)	25–30%
Equipment (reactors, solid/liquid and liquid/liquid separators, dryers, tanks, pumps. etc.)	≈20%
Building (including heating, ventilation. and air conditioning)	15–20%
Process control, instrumentation. and electrical installation	10–15%
Engineering	10–15%
Qualification and start-up	5%
Contingencies	5%

kilogram (kg^{-1}) basis typically decrease substantially [18]. The variations of the other voices are self-explanatory.

7.2   MANUFACTURING COSTS

The raw material consumption and the conversion cost are the two most important elements that establish the manufacturing cost for a particular fine chemical. The former is determined primarily by the unit consumption and the purchasing cost of the materials used; the latter, by the throughput in kilograms per day in a given production bay. A precise calculation of the conversion cost is a demanding task. Different products with widely differing throughputs (see Section 5.2) are produced in campaigns in multipurpose plants, occupying the equipment to different extents. Therefore, both the production capacity and the equipment utilization for a specific fine chemical are difficult to determine. Moreover, cost elements such as labor, capital, utilities, maintenance, waste disposal, and quality control cannot be allocated unambiguously.

A pragmatic approach for calculating the conversion cost consists in

1. Calculating the volume and time-specific standard operating cost of a plant:

Standard operating cost

$$= \frac{\text{total yearly operating cost (\$)}}{\text{operating hours (h)} \times \text{reactor volume (m}^3) \times \text{cap. utilization (\%)}}$$

Example (1):

1. Calculation of the *standard operating cost*
   - Total yearly operating cost $= \$20,000,000$
   - Operating hours per year $= 7920$ (330 days at 24 h)
   - Reactor volume $= 100\,\text{m}^3$
   - Capacity utilization $= 60\%$
   - *Available capacity* $= 475,200\,m^3 \times h/year$

   $\rightarrow$ Standard operating cost $= \$42.09\,\text{m}^{-3} \times \text{h}^{-1} = \textbf{a}$

Note: For typical operating costs in function of ownership and location of the plant and reactor size see Table 14.3.

2. Calculation of the *volume × time/output* (VTO) for a specific process step:

$$\text{VTO} = \frac{\text{total volume requirement (m}^3) \times \text{reaction time (h)}}{\text{product output (kg)} \times \text{OF}^*}$$

Example (2):

- Total volume requirement $= 9\,\text{m}^3$ (e.g., one bay with three reactors at $3\,\text{m}^3$)
- Reaction time $= 30\,\text{h}$
- Product output $= 1000\,\text{kg}$
- OF $= 0.9$

$\rightarrow$ VTO $= 0.3\,\text{m}^3 \times \text{h} \times \text{kg}^{-1} = \textbf{b}$

Note: Empirical VTO values are <0.5 for an effective 0.5–1.0 for an average and >1.0 for an ineffective process.

---

* OF = operating factor. Typically between 0.6 and 0.9. Determined by (1) length of the production campaign; (2) time lost during running-in and out (cleaning and reconfiguration); and (3) learning curve.

3. *Conversion cost* $= \mathbf{a} \times \mathbf{b} = 12.63 \, \$/kg$

   Note: In order to arrive at the full manufacturing cost, raw material costs must be added.

4. *Production capacity estimation by VTO*

   Example (3):

   - VTO $= 3 \, m^3 \times h^{-1} \times kg^{-1}/$(API, three-step synthesis)
   - Available capacity $= 475{,}200 \, m^3 \times h/year$ (see example [1])
   - Calculation: 475,200/3 $= 158{,}000 \, kg$

   $\rightarrow$*158,000 kgs of API can be manufactured in this bay.*

A first, approximate calculation can be done by a process development or pilot plant chemist on the basis of the laboratory synthesis procedure. Controlling has to be involved for a more in-depth costing, as would be required for negotiating a supply contract. The problems it has to address are how to fairly allocate costs for production capacity, which is not used. This can be due to the fact that part of a production bay is idle, because, for example, a reactor is not required for a particular process (see also Section 5.2). Also a lack of demand can be an issue. The cleaning procedure required at the end of a campaign also absorbs production capacity.

> If the campaigns are short, cleaning can be the biggest «product» in a multipurpose plant
>
> —Rolf Dach (Boehringer-Ingelheim)

If a correction for the unutilized equipment is not made, simple products, for which only part of the bay is used, for example, liquids which do not require centrifuges or dryers, can show an attractive profit margin without providing a good return for the overall investment in the multipurpose plant. For portfolio optimization, not only the profit margin per production day but also the overall income for a certain period should be considered. In other words, marketing has to find a balance between pricing or costing of individual products and overall profit optimization within specific equipment based on a planned capacity utilization taking the risk of empty capacity or unforeseen bottlenecks into account.

A particularly controversial issue is the allocation of R&D costs, as only a minority of new products or processes studied in R&D enjoy commercial success. This problem is usually disguised by not including R&D in the cost calculation of individual products, but by placing R&D in the general overhead.

An indicative cost structure for a fine chemical company is shown in Table 7.3. The operating schedule also has a significant impact on the production costs. Whereas continuous plants typically run 24 h/day, there is more freedom

TABLE 7.3   Indicative Cost Structure of a Fine Chemical Company

Cost Elements			Details	Share
Raw Materials			Inclusive solvents	30%
Conversion cost	Plant Specific	Utilities and Energy	Electric power, steam, brine	4–5%
		Plant labor	Shift and daytime work	10–15%
		Capital Cost	Depreciation and interest on capital	15%
		Plant Overhead	QC, maintenance, waste disposal, etc.	10%
		Research & Development	Inclusive pilot plant	8%
	Marketing & Sales		Inclusive promotion	5%
	General Overhead		Administrative services	15%
Total			Exclusive taxes	100%

in establishing operating schedules for multipurpose plants. Depending on the workload and the flexibility of the workforce, schedules can be adjusted as needed. Some schedules still include only a one-shift or a two-shift operation (e.g., 8 h or 16 h per day for 5 days a week). In this case, frequently, some minimum activity is maintained during the night, such as supervision of reflux reactions, solvent distillations, or product drying. Nowadays, a full 7-day/week operation, consisting of four or five shift teams, each working 8 h per day, has become the standard. In terms of production costs, this is the most advantageous scheme. Higher salaries for night work are more than offset by better fixed cost absorption. Also, only part of the workforce has to adhere to this scheme.

Manufacturing costs usually are reported on a per kilogram product basis. For the purpose of benchmarking (both internal and external), the *volume × time/output* (VTO), as mentioned above, is a useful aid. *Campaign reports* are a valuable tool for monitoring the plant performance for a specific product. During budgeting, standard costs for a production campaign of a particular fine chemical are determined on the basis of past experience. The actual results of the next campaign are then compared with the standard.

When the attractiveness of new products is evaluated, either for submitting an offer or for inclusion in the R&D program, manufacturing costs have to be estimated on the basis of a laboratory synthesis procedure. This is best done by breaking down the process into unit operations, the standard costs of which have been determined previously. Care has to be taken to estimate the time required for each step of a process. Thus a liquid–liquid extraction can take more time than the chemical reaction. The capability of a fine chemical company to make dependable manufacturing cost forecasts is a distinct competitive advantage.

For a fine chemical company committed to sustainable development, the manufacturing cost expressed in $/kg as such must not be the only criterion for the selection of a production process for a given product. In order to assess the environment impact, several metrics have been developed. Examples are the *Atom Economy*, indicating which proportion of the reactants remain in the final product, the *Carbon Efficiency*, measuring the amount of C in product versus total C in reactants, Roger Sheldon's *Environmental (E)-Factor*, indicating the total quantity of waste produced per kilogram of product. For fine chemicals, the number is typically in the range of 5–50.

CHAPTER 8

Management Aspects

Successfully managing a fine chemical company presents a greater challenge than running a more conventional manufacturing company. The fine chemical industry is relatively small, very competitive, complex, and capital-intensive. The transactions are of the "b2B" type, whereby "B" stands for powerful life science companies in most cases. The single major vulnerability is the lack of "product equity." A fine chemical producer simply does not control the destiny of the goods or services sold. Managing the risk–reward balance therefore means a great challenge for both the organization, systems, and business processes in general and the chief executive officer (CEO) in particular. Apart from the usual entrepreneurial capabilities of "competency and commitment," the CEO should possess a combination of a strong chemical background (i.e., an academic degree in a chemistry or chemical engineering-related subject plus substantial industry experience, preferably in a research laboratory and manufacturing plant), a view of the pragmatically feasible, and business acumen. As part of their "natural latent abilities," CEOs must be good communicators and negotiators will all types of stakeholders, extending from customers, employees, environmentalists, suppliers, investors to regulatory and government officers.

> Because of our crisis period, we are paranoid about the concentration of risk— whether it's customers, technology, markets, plants, [environment] you name it ... and created the position of "chief risk officer" at the Management Committee level
>
> *Source*: Lonza

Considerable experience supports the notion that in complex situations such as this, expectation is realized only by appropriate measurement, normally represented by measures of performance (MOPs). Measurement is, however, fraught with danger as it will have an undoubted effect on behavior, and it is therefore crucial that the right things are measured so that the organization

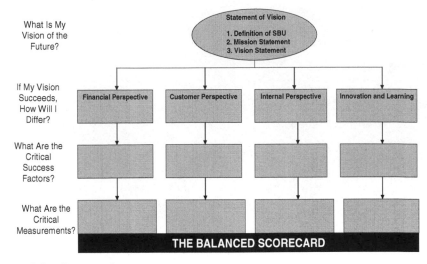

Figure 8.1 Connected measurement.
Source: Norton and Kaplan, *HBR* [21].

does both "the right thing" and "the thing right." Thoughtful consideration of the difference between drivers and outcomes is essential, as is a clear understanding of responsibility and accountability. Another important consideration is the so-called "line of sight" between individuals who can effect performance that impacts a driver and the consequential outcome.

Kaplan and Norton at Harvard Business School recognized many of these issues and through multi-company research developed something of a top-down solution called the *balanced scorecard* [21]. During the 1990s, this concept was constantly developed with papers published in *Harvard Business Review*. A simple representation is shown in Figure 8.1 and depicts good connectivity and alignment between measures and the vision and strategy of the company. It is still necessary, however, to differentiate between measures of outcomes that act as *milestones* to demonstrate progress with the strategy as opposed to measures of *drivers* that will result in those outcomes.

A good example is *working capital reduction* measured by *inventory turn*. In the short term, simply measuring this may result in initial improvement that helps the business, but without a systematic improvement program aimed at the drivers, there may not be a sustainable benefit. The required drivers would be campaign length and changeover times in a typical multipurpose plant with multiple sequential product campaigns (see also Section 5.2).

A more recent development is to use *dashboards* as a subset of a balanced scorecard or measures at the next level down [22]. These dashboards allow getting a concise insight in how well an organization is performing overall. They are now widely used in management information packs for key executive team meetings and also locally on the shop floor for further improving transparency and organizational connectivity between different functions.

Each of these key issues is now explored in more detail in the two next sections.

8.1 RISK/REWARD PROFILE

The fine chemical industry embodies a complex combination of physical assets, technologies, know-how, and intellectual property (IP) that are geared to commercial process development and manufacturing. Since the mid-1990s, the marketplace has become much more competitive with a proliferation of independent players and, during the last decade, unstoppable, rapidly emerging strength from low-cost economies (see Chapter 14).

Some of the characteristics of the industry are

Overall

- High fixed capital intensity
- High working capital
- High fixed costs
- High-quality standards and demanding regulations
- Long supply chains
- Short product life cycles
- Considerable yield and material waste

Custom Manufacturing (CM)

- Existing customers have a tendency to "pigeon-hole" suppliers in a way that can exclude them from consideration for a proportion of new projects.
- Potential customers have "preferred supplier" lists excluding unlisted companies from the best projects.
- Existence of surplus capacity tends to be the norm, and the resulting significant cost burden is rarely properly recompensed by customers.
- Intense pressure to reduce costs does not allow producers adequate leeway to improve process economics.
- Customers usually deny their suppliers any economic return for process improvements; often, such changes are even prohibited. Margins only decrease with time.
- When the drug resulting from an outsourced fine chemical becomes a commercial success, the customer will often take part or all of the production back in-house.

TABLE 8.1   Fine Chemical Industry Benchmark Data

Benchmark Data	% of Sales
Fixed assets	90
Working capital	35
Net Operating Assets (NOA)	125
Return on Net Operating Assets (RONOA)	7½
Depreciation	8
Direct labor	12
Research & development	6
QA/QC	4
Marketing & sales	3
Gross margin	55
Return on sales (ROS)	9

Source: Author's estimates.

For quantitative data, see Table 8.1. In the market, considerable risk exists because of the structure of the customer base, primarily the pharmaceutical industry:

- High and increasing product attrition within customer portfolios
- Strong consolidation of customer base (14 major pharmaceutical companies account for more than 50% of global demand)
- Strong purchasing power (customers usually are much larger than suppliers)
- Competition from backward-integrated customers who have a major share of capability and capacity within the industry
- Intellectual Property rights stay with customer

In addition to this, the fine chemical industry does not offer genuine differentiation and responsiveness compared to other industries. Actually, the announcements made by the companies look strikingly similar:

- State-of-the art facilities
- Versatile equipment
- Room for expansion
- Large toolbox of technologies
- Creative process development
- Seamless transition from laboratory to plant
- Full compliance with current Good Manufacturing Practice (cGMP)
- FDA-approved plants
- Highly efficient safety, health, and environment (SHE) management

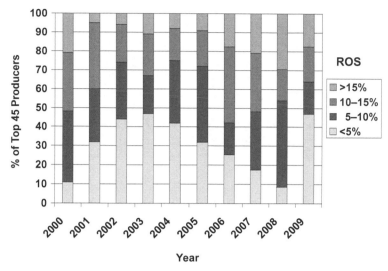

Figure 8.2   Profitability in the fine chemical industry.
Source: Ramakers fine chemical benchmarking database.

The consequences are high risks on both cost and demand sides. The impact of these industry characteristics on the return on sales (ROS) is shown in Figure 8.2. These data come from a 45-company cohort and are believed to be representative of the Western fine chemical industry. As can be seen, profitability was dramatically affected as the number of new chemical entities (NCEs) declined from the end of the 1990s onward. By 2002, over 50% of the sample had profitability below the average cost of capital and would have struggled to maintain investment in their businesses. The 2003 data show strong companies getting stronger and weak companies further weakening. This finding apparently contradicts the preceding statement, namely, that there is no differentiation among the industry. The explanation is that there actually is little or no differentiation in hard facts, such as types of equipment used, but a lot in soft issues, such the reputation, built on a track record of successfully realized CM projects and FDA inspections. Other soft factors such as flexibility, agility, and ease of doing business have an increasing impact on differentiation. These often depend on a strong balance sheet enabling the supplier to be less rigidly contractual on all matters while maintaining a high level of financial robustness. The global industry improvement through the period 2004–2008 positively impacted on the profitability of the fine chemical industry. It is interesting to contrast peak year 2008 with the previous peak in 2000. While the proportion making more than 10% ROS has returned to a very similar level of approximately 50%, the proportion making more than 15% return has increased substantially from 20% to 30% of the population, confirming the notion that the strong are getting stronger.

The financial crisis of late 2008 has had a dramatic impact with almost 50% of the group having profitability less than 5% in 2009. This is more than a fivefold increase in unacceptable profitability happening over a period of only 12 months and is a much more dramatic collapse than the early part of the decade. This is primarily because the earlier profit collapse was driven largely by an excess of supply capacity coinciding with a reduction of demand within the industry rather than a worldwide recession.

Since 2008, however, the global financial crisis has triggered a recession in most developed economies which has also coincided with a period of substantial industry restructuring among the major multinational pharmaceutical companies which are now preparing for the patent expiry cliff in 2012.

This "perfect storm" of events has created a very adverse environment for contract development and manufacturing, and while there is some sign of this bottoming out there, is no indication of recovery.

With such a disparity between the best (ROS >15%) and worst (ROS <5%, sometimes negative) performers, a "survival of the fittest" condition has been established (see Chapter 20).

8.2   PERFORMANCE METRICS AND BENCHMARKING

In order to determine the strengths and weaknesses of a particular fine chemical company, it is worthwhile to gather performance benchmarks both within the company and with important competitors. There are, however, a number of potential pitfalls in collating data from both inside and outside the industry. On the inside there may be significant subsectors, sometimes referred to as *segments*, where the offering and business model can vary substantially so that taking over data unbiased can be very misleading. On an external comparison, getting appropriate industry "peer groups" is important. Judging the balance of technology, development, manufacturing, and service to get meaningful comparators is complex and difficult.

Benchmark data collected in a study of a representative number of European and U.S. Fine Chemical companies over the 1998–2003 period are listed in Table 8.1. The average return on net operating assets (RONOA) of the sample was 9%. The last two figures exemplify the capital intensity of CM-biased fine chemical companies. In order to achieve a good return of capital employed, a healthy margin is mandatory. A more standard product-dominated business portfolio would not only have significantly lower capital intensity but also lower margins at RONOA and ROS levels.

A better performance would improve these measures, but these parameters are largely measures of outcomes, not measures of performance on drivers.

The notion of a *balanced scorecard* linking strategic vision to operational delivery and measures of performance was developed by the Nolan Norton Institute (research arm of KPMG). David Norton and Robert Kaplan undertook a multi-company study to help develop the performance model. Norton's network in the consulting arena helped establish some momentum with many companies in a diverse array of industries embracing the methodology. The system developed and senior executives started to use the scorecard to develop company strategy, not just measure performance. The most complete collection of reports is reviewed in Reference [21]. The five-volume set includes balanced scorecard basics, human capital, mapping strategy, and both the 2004 and 2005 editions of the Balanced Scorecard Report Hall of Fame.

Figure 8.1 provides a vertically connected tool for aligning the plethora of financial, operational, and development activities to the strategy and the vision. This facilitates an organizational alignment whereby employees can relate their activities and measures to the delivery of the strategy and the consequent success of their corporation or enterprise. In addition, the scorecard should be read right to left along the bottom as the learning, innovation, and growth perspective should feed and develop the internal perspective, which should be responding to customer needs. The combination of this sequence of activity will result in the financial outcome. Although the concept has many obvious attractions, care is required to ensure that on implementation at a specific company, the correct measures are selected, and individuals and teams have a clear "line of sight" on how their actions and measures affect the implementation of the strategy and the consequential financial results.

The constant advance of regulation and compliance coupled with a strong wish to self-regulate in the chemical industry resulted in initiatives such as responsible care. Voluntary external auditing and accreditation was necessary to establish standards and measure compliance. Thus, as the 1990s proceeded, accreditation of ISO 9000, ISO 9001, ISO 9002, and ISO 14000 became increasingly popular. Such a system forces a company to define its organization and business processes with the commensurate identification of measures of performance (see following section).

This methodology inevitably leads to a resultant company performance scorecard, an example of which is shown in Appendix A.4. In the fine chemical arena, regulatory bodies such as the American Food and Drug Authority (FDA), the Chinese State Food and Drug Administration (SDA), and the European Medicines Agency (EMEA), which focus on establishing and ensuring cGMP, can have a big impact on the measures chosen. Thus, measures such as "really right first time" or "process deviations" become very important. Systems such as ISO 9000 bring a structure to what is to be welcomed, and such transparency and measurements will definitely yield benefit, but it can sometimes institutionalize an existing organization that in reality is inappropriate for the strategy and business model.

This methodology has value in providing the necessary guidance and roadmap for individuals in the company. The ISO concept of a complete inventory of measures should be used as a backup for choosing the most appropriate measures for any given function and time period. As the business develops and a large inventory of MOPs is maintained, it will be appropriate to focus during each annual review for each employee on a small number of key measures of performance (drivers and outcomes) that will be prioritized for the whole organization in that particular budget period. As a result, all employees can identify with at least one vital measure and have a clear "line of sight" on the impact of their individual or team contribution. Such measures must be connected to a formalized improvement program that ensures delivery and sustainability. The basis to develop the system is clearly defined responsibilities, accountabilities, and line of sight. Last but not least, it will be management's task to clearly communicate progress and success.

8.3 ORGANIZATION

A manufacturing company can basically be organized according to assets, geographic regions, or businesses. More than in other industries, the destiny of the fine chemical/custom manufacture industry depends on the individual businesses, respectively their customers. The latter decide which product is made by whom, in which quantity, and by what process, and they themselves are subject to verdicts of the regulatory bodies. In due consideration of the customer's impact, the businesses come forward as main organizational element. This is reflected by the basic structure outlined in Figure 8.3.

The structure is appropriate for companies with more than 250 employees. It allows for the delegation of the profit and loss (P&L) responsibility from top to middle management, which has the best handle on all the drivers of the

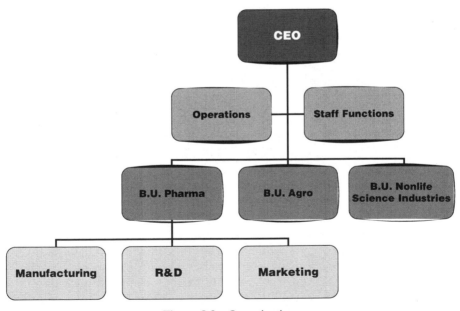

Figure 8.3 Organization.

business. "Operations" is in charge of enterprise resource planning, supply chain management, overall equipment effectiveness (OEE), plant infrastructure (e.g., utilities, maintenance, warehouses, waste disposal), QA/QC, and SHE. "Staff functions" comprise human resources, finance, legal, and general administration. "Manufacturing" refers *en principe* to the production trains in which the products for the individual business units are produced. In the routine of an enterprise, the allocation of manufacturing assets can constitute a pitfall of this business-oriented organizational structure. Given the multipurpose design of the plants, the same production train actually may produce for more than one business unit. This problem can be solved by applying the "major user" principle. For the R&D function, a flat structure, which allows for the rapid formation and dissolution of project teams, should be chosen. For a description of the "marketing" function, see Section 12.1. For smaller fine chemical companies, a subdivision of manufacturing is not advisable. In this case, the business units are limited to comprise R&D and marketing.

In the description of both the detailed skeleton and of individual functions, ISO systems and cGMP regulations will have to be observed (see previous chapter). The assignment of the particular functions can cause internal friction because the egos of individuals are at stake. An example in this case is the pilot plant, which can report to either R&D or manufacturing. The fact that there are more—and more demanding—technology transfers from laboratory to pilot plant than from pilot plant to industrial scale speaks in favor of the first option. The fact that both the pilot and industrial plants use the same site infrastructure (utilities, warehouses, maintenance, internal transports, etc.) favors the subordination to manufacturing. An authoritarian decision by the CEO can avoid a lot of wasted time in such situations.

A classical pyramidal organization model with strong functions may embody a "silo" condition that prevents effective team engagement and overall alignment; thus, ensuring that the company has the correctly engineered organization as a prerequisite for success and measurement alone will be insufficient.

Once the detailed organizational structure is in place, job descriptions (e.g., see Appendix A.5), which should be an integral part of the organization, are prepared, and individuals are assigned to the various functions, fundamental changes should be avoided. Actually, many companies suffer from the "organization carousel syndrome"; after drawbacks of the actual organization become apparent, a paradigm shift is believed to be necessary in the (vain) attempt to establish the perfect organization. A vicious cycle typically starts with the appointment of a new CEO, or the creation of an overseas subsidiary, which calls for a geography-based organization. After the disadvantages of the new model have become apparent (e.g., because the plants in the United States and EU produce for the same customers), an asset-based organizational structure is drawn, which does not meet the ideal of the perfect organization, either. This becomes apparent when customers report that they have been visited by marketing people from different asset owners. In the end, the original scheme, perhaps under a different name, is readopted. One complete turn of the organization carousel typically lasts for about15–20 years, depending on the fluctuation rate in top management. Overall, a too hectic (re-) organization activity is counterproductive. Reviews have confirmed again and again that the hidden costs of reorganizations, such as widespread uncertainty, demotivation, and loss of institutional memory are much higher than the advantages of new structures.

■■■■■■ BIBLIOGRAPHY

CITED PUBLICATIONS

1. *Encyclopedia of Reagents for Organic Synthesis*, 8-vol. set, Prof Leo A. Paquette (editor-in-chief), Wiley, New York, 1995.
2. *Houben-Weyl, Methods of Organic Chemistry*, Thieme, Stuttgart, Germany, 2001.
3. Process Directory (section), *Informex Show Guide*, 2005, pp. 123–167.
4. A. Thatcher, Handle with care, *Chemical & Engineering News*, March 16, 2009, pp. 12–19.
5. C. Chassin and P. Pollak, Outlook for chemical and biochemical manufacturing, *PharmaChem* 1–2, 135–136 (January/February 2004).
6. J. Rauch, *Mehrprodukteanlagen*, Wiley-VCH Verlag, Weinheim, Germany, 1998.
7. R. Hansen, *Overall Equipment Effectiveness (OEE)*, Industrial Press, New York, USA, 2005, ISBN (978-0-8311-3237-8);
   7a. 21 Code of Federal Regulations; Guidance for Industry, Part 11, Electronic Records; Electronic Signatures, Scope and Application, US Department of Health and Human Services, Food and Drug Administration (1 CFR part 11), August, 2003.
8. *European Roadmap for Process Intensification*. http://www.creative-energy.org. The Netherlands, 2008.
9. V. Hessel, B. Loeb et al., Process intensification, in M. V. Koch, K. M. VandenBussche, and R. W. Chrisman (eds.), *Micro Instrumentation for High Throughput Experimentation and Process Intensification*, Wiley-VCh, Weinheim, Germany, 2007, pp. 85–129.
10. FDA report, Pharmaceutical Quality for the 21st Century: A Risk-Based Approach. http://www.fda.com (2007).
11. F3Factory Consortium. http://www.f3factory.com (2009).
12. ICH Q8, *Good Manufacturing Practice for Active Pharmaceutical Ingredients*. http://www.fda.gov.
13. 21 Code of Federal Regulations; *Part 210*, Current Good Manufacturing Practice in Manufacturing, Processing, Packing or Holding of Drugs, US Department of Health and Human Services, Food and Drug Administration (21 CFR Part 11), US Department of Health and Human Services, Food and Drug Administration, August 2001, ICH. http://www.fda.gov.

Fine Chemicals: The Industry and the Business, Second Edition, by Peter Pollak
Copyright © 2011 John Wiley & Sons, Inc.

14. *European Pharmacopoeia*, 5th ed., EDQM, European Pharmacopoeia, Council of Europe, B.P. 907, F-67029, Strasbourg, France, July 2004.

15. *Pharmacopoeia Internationalis*, 3rd ed., World Health Organization, Geneva, Switzerland, 2003,

16. *Food Chemicals Codex (FCC)*, 5th ed., Institute of Medicine, Washington, DC, 2003,

17. *Reagent Chemicals—ACS Specifications*, 9th ed., American Chemical Society, Washington, DC, September 1999.

18. B. Gujral, C. F. Stanfield, and D. Rufino, Focus on PAT, Supplement to *Chimica Oggi/Chemistry Today* 26 (**1**), 10–11 (2008).

19. C. Rosas, From the bench to the pilot plant and beyond, in S. H. Nusim (ed.), *Active Pharmaceutical Ingredients*, Taylor & Francis, Boca Raton, FL, 2005, pp. 41–42.

20. P. Romagnoli, *Business Development in GMP Fine Chemicals*, Romagnoli Consulting s.n.c., Milan, 2003, pp. 297–304.

21. R. S. Kaplan and D. P. Norton, *Breakthrough Results with the Balanced Scorecard, A Report Collection*, 2nd ed., Harvard Business Publishing, Boston MA, USA, 2007, Boston, MA, 2007.

22. W. Wayne, *Eckerson, Performance Dashboards, Measuring, Monitoring and Managing Your Business*, John Wiley & Sons, Hoboken, NJ, 2006.

FURTHER READING

R. Bryant, *Pharmaceutical Fine Chemicals: Global Perspectives*, Informa Publishing Group, London, 2000.

A. Cybulski et al., *Fine Chemicals Manufacture—Technology and Engineering*, Elsevier, 2001, 564 pages.

M. C. Flickiger, ed., *Encyclopedia of Industrial Biotechnology* (7-vol. set), J. Wiley & Sons, Hoboken, NJ, 2010.

A. Kleemann and J. Engel, *Pharmaceutical Substances*, 4th ed., Thieme-Verlag, Stuttgart, New York, 2001.

H.-R. Marti and J. S. Siegel, eds., Process development of active pharmaceutical intermediates following a development cascade, *Chimia* 60 (**9**), 523–629 (2006). http://www.chimia.ch.

The Merck Index, 14th ed., Merck & Co., Whitehouse Station, NJ, 2007.

S. H. Nusim, ed., *Active Pharmaceutical Ingredients: Development, Manufacturing and Regulation*, Taylor & Francis, Boca Raton, FL, 2005.

P. Pollak and E. Habegger, Fine chemicals, in *Kirk-Othmer Encyclopedia of Chemical Technology*, 5th ed., vol. 11, pp. 423–447, Wiley, New York, 2005.

C. S. Rao, *The Chemistry of Process Development in Fine Chemical and Pharmaceutical Industry*, Asian Books Private Ltd., 2004, 1311 pages—ISBN 81-86299-50-5.

PART II

THE BUSINESS*

---

* In this part, a static assessment of the fine chemicals business is made. A dynamic appraisal is made in Part III, OUTLOOK.

CHAPTER 9

Market Size and Structure

9.1 FINE CHEMICAL MARKET SIZE

As stated in Chapter 1, fine chemicals account for only for about 4% of the universe of chemicals. The latter, valued at $2500 billion, is dominated mainly by oil-, gas-, and mineral-derived commodities on the one hand, and a large variety of specialty chemicals at the interface between industry and the public on the other hand (see Table 9.1).

All fine chemicals are used for making specialty chemicals, either by direct formulation or after chemical/biochemical transformation of intermediates to active substances. Specialty chemicals are solid (e.g., tablets) or liquid (e.g., solutions) mixtures of commodities or fine chemicals and exhibit specific properties. They are sold on the basis of "what they can do" (e.g., protect the skin against ultraviolet radiation), rather than on "what they are" (e.g., molecular structure 2-ethyl-hexylmethoxycinnamate). Within specialties, pharmaceuticals and other life science products use the largest amount of fine chemicals. They are described in detail in Sections 11.1–11.3. Uses of fine chemicals outside life sciences are discussed in Section 11.4.

There are three main reasons why it is not possible to exactly determine the size of the fine chemical market:

- The definition of fine chemicals is not very accurate (see Section 1.1 text and Figure 1.1). Adopting the "$10 per kilogram" as the threshold for distinguishing between commodities and fine chemicals, it cuts right into several otherwise homogeneous product classes. Thus, within vitamins, niacin (vitamin B_3) is classified as a commodity, and biotin (vitamin H) and vitamins D and E as fine chemicals. The same occurs within amino acids, where D,L-methionine and L-lysine are commodities, whereas L-proline and L-tryptophane are fine chemicals. This somewhat disturbing inconsistency is also justified by the fact that the products at the low end of the price range of these classes are produced in dedicated plants in

Fine Chemicals: The Industry and the Business, Second Edition, by Peter Pollak
Copyright © 2011 John Wiley & Sons, Inc.

TABLE 9.1   Structure of the $2500 Billion Global Chemical Market

Specialties ≈55% ($1400 billion)	Additives and catalysts Biocides Dyestuffs and pigments Electronic chemicals Flavors and fragrances Food and feed additives Household and personal care **Life science products** Specialty polymers	✓**Pharmaceuticals** Agrochemicals ↖Veterinary drugs
Fine Chemicals ≈3% ($90 billion)	Standard and advanced intermediates Active substances for specialties, especially drugs (human and veterinary) and pesticides	
Commodities ≈40% ($1000 billion)	Petrochemicals Plastics and synthetic rubber Synthetic fibers Fertilizers Other inorganic chemicals	

↑ Added value →

quantities exceeding 10,000—in some cases even 100,000—tons per year. These are typical characteristics of commodities.

- Many fine chemicals are not traded, but are produced in-house by diversified chemical and life science companies. One must distinguish, therefore, between "merchant value" and "captive value" of fine chemicals (FCs in the following equation). The former is the product of *traded* quantities times the unit sales prices; the latter, the product of *produced* quantities times (virtual) unit prices of captively manufactured products. The total value is the sum of both merchant and production values:

$$\text{Total value: } \sum (p_{1-m} \times q_{1-m}) + \sum (P_{1-n} \times Q_{1-n})$$

p = unit price of traded fine chemicals q = volume of traded fine chemicals	>merchant value	>total value
P = (virtual) unit price of captively produced FCs Q = volume of captively produced FCs	>captive value	

One can also reason that the merchant value corresponds to the actual aggregated sales revenues of the fine chemical industry, whereas the total value is the theoretically achievable revenue, if all fine chemical requirements were outsourced.

• The official trade and custom statistics are not broken down to fine chemicals. The U.S. Bureau of the Census classification, for instance, uses the following main product categories: "organic chemicals," "inorganic chemicals," "plastics," "fertilizers," "pharmaceuticals," "cosmetics," "dyes and colorants," and "other."

In order to arrive at a realistic figure for the size of the fine chemical business, a top-down approach has been adopted. For the total value, it is based on three key figures, namely, the total sales of a category of specialty chemicals (e.g., pharmaceuticals) and the percentage share of active ingredients. For determination of the merchant market, the latter figure is multiplied by the share of outsourcing. As the total sales figures for major specialties are published and the share of active ingredients in the specialties is known approximately, the total value of fine chemicals can be estimated with reasonable accuracy. As no reliable data exist on the degree of outsourcing, estimation of the size of the merchant market is more speculative.

9.2 MARKET BREAKDOWN BY MAJOR APPLICATIONS

In Table 9.2, the approximately $80 billion fine chemical market is subdivided into major applications according to their relevance, namely, fine chemicals for pharmaceuticals, agrochemicals, and specialty chemicals outside life sciences. Furthermore, a distinction is made between captive (in-house) production and merchant market.

Pharmaceutical fine chemicals (PFCs) account for more than two-thirds of the total value of fine chemicals. The methodology for arriving at the figure of $55 billion for the PFCs is shown in Table 9.3. An average percentage of 6.5% (5% for proprietary drugs, 15–20% for generics) has been applied for the

Table 9.2   Market Breakdown by Major Applications

		Size ($ billion)		
		Total a.i.	Captive	Merchant
Life sciences	Pharmaceuticals	55	32	23
	Agrochemicals	15	11	4
Various specialty chemicals		15	10	5
Total fine chemical industry		**85**	**53**	**32**

a.i., active ingredient.

TABLE 9.3 **Structure of the Pharmaceutical Fine Chemicals Market**

Formulated Drugs		Pharmaceutical Fine Chemicals[a]		
Total Pharma Market 2010[E] 100%/$835 b	↘ Share of proprietary drugs 88%/$735 b	↘ Share of active ingredients[b] 5%/$37 b	↘ Share of outsourcing 33%/$12 b
	↘ Share of nonproprietary drugs 12%/$100 b	↘ Share of active ingredients[b] 15–20%/$18 b ↓ Total active ingredients $55 b	↘ Share of outsourcing 60%/$11 b ↓ Total share of merchant market 42%/$23 b[c]	↘

[a] Accuracy of the numbers ± 25%.
[b] APIs + advanced intermediates.
[c] For a further breakdown of the markets according to types of companies and drug development stage, see Table 12.6 in Section 12.3.

PFCs' share. Out of the total PFC value of $55 billion, about $16 billion (~30%) is traded, and $39 billion is the production value of the pharma industry's in-house production. The accuracy of these figures is ~20–25%. Within life science products, fine chemicals for agro, and—at a distance—for veterinary drugs follow in importance.

Global sales of proprietary drugs are estimated at $735 billion in 2010, or almost 90% of the total pharma market. With a share of 45%, the United States has by far the single largest market. Global sales of generics are about $100 billion, or just over 10% of the total pharma market. Obviously, the difference in production volumes is smaller.

For agrochemicals, an incidence of the active fine chemical ingredients of 25% has been assumed, resulting in a total value of $10 billion out of $40 billion for the global pesticide market. The percentage is much higher than for pharmaceuticals, especially proprietary drugs, because cost/benefit considerations are more important in the agrochemical industry. A farmer uses a pesticide only if such use provides an economic benefit, whereas a patient with a reimbursable prescription does not care much about the cost of the medicine.

For the value of fine chemicals used in specialty chemicals outside life sciences, only a best guess is possible. Issues such as "at what stage of the supply chain the end price is determined (wholesaler, supermarket, or specialty boutique)," "what is the exact concentration of the embedded fine chemical"

TABLE 9.4   Embedded Value of Fine Chemicals in Specialty Chemicals

Product	L-lysine	Aerogard	Listerine	SuperGlue
Application	Vitamin supplement	Insect repellent	Mouthwash	Adhesive
Wholesale price	$10.00 *0.1 kg*	$6.00 *0.15 kg*	$5.00 *1.5 kg*	$1.50 *3 g*
package size Unit price	$100/kg	$40/kg	$3.30/kg	$500/kg
Key ingredient	L-lysine	Diethyl toluamide	Ethanol	α-cyano acrylate
concentration	*>95%*	*19%*	*27%*	*>95%*
share of price	**≈5%**	**≈4%**	**≈3%**	**≈4%**
Other ingredients		di-n-propyl isocinchomerate, N-octyl bicyclo-heptenedicar-boxamide	Eucalyptol, menthol, methyl salicylate, thymol	

(strengths of sunscreen formulations vary by a factor of 10), or "how do formulations from different companies vary" are difficult to address with mathematical precision. Furthermore, many specialty chemicals contain more than one active ingredient; laundry detergents, for instance, contain zeolithes (15–30%), tensioactives (5–15%), anti-redeposition agents, foam regulators, optical brighteners, perfume, proteoloytic enzymes, polycarboxylates, and other compounds (<5%). For an order-of-magnitude estimate, four household specialty chemicals have been selected randomly and analyzed (see Table 9.4). Although the weight shares of the active ingredients vary widely (viz., between 19% and >95%), the variation of the embedded values of the active ingredients all range between 3% and 5%. It can be assumed, therefore, that the total value of active ingredients in specialty chemicals corresponds to 4% of the global sales revenue of formulated specialty chemicals, which amounts to $775 billion [$1650 billion of which $875 billion for (pharma + agro)]. Assuming furthermore that half of the active ingredients are commodities, this translates into a market value of about $15 billion.

As the leading specialty chemical companies, Akzo Nobel, Dow, Du Pont, Evonik, Chemtura, and Mitsubishi are backward integrated, the share of in-house production is estimated at 75%, leaving a merchant market of approximately $5 billion (see Table 9.2).

CHAPTER 10

The Business Condition

The fine chemical industry has undergone three boom and two bust phases during its almost 30 years of existence. The first boom, which led to the creation of the industry, occurred in the late 1970s, when the overwhelming success of the histamine H_2 receptor antagonists Tagamet (cimetidine) and Zantac (ranitidine hydrochloride) created a strong demand, which could not be covered by in-house production, for the active ingredients and preceding steps. These anti-stomach ulcer drugs virtually eliminated the need for stomach surgery. The second boom took place in the late 1990s, when both high-dosage anti-AIDS drugs and COX-2 inhibitors required much multipurpose manufacturing capacity and gave a big boost to custom manufacturing. The most recent—minor—boom is associated with stockpiling of GlaxoSmithKline's Relenza (zanamivir) and Roche's Tamiflu (oseltamivir phosphate) by many countries in order to prepare for a possible avian flu epidemic. For the procurement of the latter, Roche had established a global network including several Roche sites and more than 15 external contractors in 9 countries.

After the second boom phase, which had seen double-digit growth rates for the fine chemical business and was described as "irrational exuberance" of the 1990s, the industry suffered a first bust after the turn of the century. As a result of the ruinous mergers and acquisitions (M&A) activity, several billion dollars of shareholder value were destroyed. Two more downturns followed in 2004–2005 and 2008–2009. The year 2005 can be considered the year of the truth for the European fine chemical industry. For the first time, several leading companies admitted unsatisfactory performance. Stefan Borgas, CEO of Lonza said: "Lonza is the only company that made money in custom manufacturing over the past two years." Surprisingly, the main cause for the 2008–2009 slump had not been general recessions at the customers' end, but slowdowns of the growth and, even more so, inventory adjustments by the pharma industry. They resulted in postponements or cancellations of orders. The unfavorable development was in sharp contrast to the very optimistic growth forecasts, which many fine chemical companies had announced. They had been based on equally promising sector reports from investment banks, which in turn had

Fine Chemicals: The Industry and the Business, Second Edition, by Peter Pollak
Copyright © 2011 John Wiley & Sons, Inc.

TABLE 10.1   Profitability of the Western Fine Chemical Industry

	Business Condition		
	"Boom"	"Bust"	Average 2000–2009
EBITDA/Sales	~20%	~10%	~15%
EBIT/Sales	~10–13%	~5%	~7–8%

evolved from forward projections of the preceding boom period. In most cases, these projections have been missed by a large margin.

The profitability of the fine chemical industry followed the ups and downs of the business condition (see Table 10.1). While the average earnings before interests, taxes, amortization, and depreciation (EBITDA)/sales and earning before interest and taxes (EBIT)/sales of representative companies, resp. divisions were 15% and 7.5%, respectively, in the period 2000–2009, the numbers were 20% and 10–13% in the boom, and 10% and 5% in the bust periods. The factor 2 between the high and low number reflects the sensitivity of the industry to variations in capacity utilization. It was also negatively affected by capacity expansions, which had been put in place in anticipation of a steadily growing demand. In order to avoid plant closures, Western fine chemical companies accepted prices that covered only variable costs, in some cases even only part of them. Last but not least, the currency exposure is another element of concern. Most costs incur in euros, whereas sales are in U.S. dollars. All in all, the average Western fine chemical firms have been making a return below the cost of capital that they are not reinvestment grade.

10.1   OFFER

The sharp increase in the fine chemical industry's production capacity—usually expressed in "m^3 cGMP reactor volume"—is a result of three main developments, namely, the upgrading of idle dyestuff intermediate plants, the acquisition of active pharmaceutical ingredient (API) manufacturing sites from pharma companies, and—most importantly—the new capacity, which has been built mainly in Asian countries. The number of exhibitors at the world's most important fine chemical trade fair, the CPhI (Chemical and Pharmaceutical Ingredients, see Appendix A.1), is a valid qualitative indicator for the expansion of the industry. A mere 16 companies participated at the first CPhI, which was held in 1990. From then on, about 75 additional exhibitors registered every year, to arrive at 1530 in 2009. Whereas there were only a number of contract research organizations (CROs), which joined from the Western hemisphere lately, the major growth came from Asia. Exhibitors from China, India, South-Korea, and Taiwan accounted for more than 40% of the total in 2009!

For a comparison of the development of fine chemical sales between industrialized and developing countries over the past 9 years, 4 Western and 4 Indian

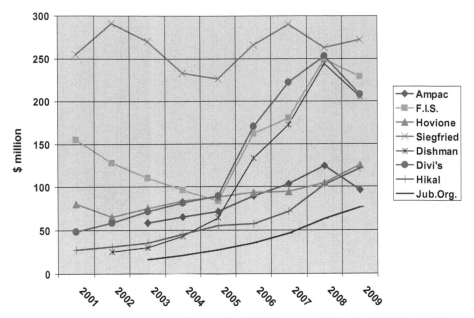

Figure 10.1   Sales development of Indian and European fine chemical companies.

fine chemical companies each have been selected. The Westerners are Ampac F.C., USA; F.I.S., Italy; Hovione, Portugal and Siegfried, Switzerland (see Figure 10.1). The Indians are Dishman, Divi's Laboratories, Hikal, and Jubilant Organosys' Pharma and Life Science Products & Services (PLSPS) Division. The aggregated sales increases of the two groups were 6% and about 25% p.a., respectively. With a growth rate of 16% p.a., family owned F.I.S., Italy, was the best Western performer. At the beginning of the period under review, all Indian companies had annual sales below $50 million. In fiscal year 2009/2010, Dishman and Divi's each generated just over $200 million, and Hikal followed suit with $120 million. Whereas sales and profits of the Westerners went through a trough in 2005, the Indians never suffered from a decline in revenues.

It is worthwhile to compare these growth rates with those of the main customer base. The sales of the pharmaceutical industry as a whole expanded by 7.5% p.a. in the same period (from $520 billion in 2003 to $805 billion in 2009), and those of the generics industry (primarily the sponsors of the Indian fine chemical companies) by 15% p.a. Thus, the sales growth of the Western companies is lower than the one of their customers, whereas the Indian ones expanded much more rapidly. The conclusion is that the Indians succeeded in cannibalizing market share of their less successful competitors, most likely mainly Western fine chemical companies.

The new players in Asia take a proactive approach and expand their production capacity ahead of demand. This allows them on the one hand to realize big new customer projects rapidly. On the other hand, their volume/ time output (see Section 7.2 has not yet reached Western standards. Divi's Laboratories, India, established in 1990, is an example of a company that has installed a very impressive production capacity. At its two large sites in Hyderabad and Vishakhapatnam (in Andhra Pradesh), the company has put on-stream multipurpose GMP fine chemical plants with reactor volumes 1293 m^3 and 1637 m^3, respectively, adding up to close to 3000 m^3, far more than any Western fine chemical company. However, their sales of $185 million (2009/2010 Estimate) translate in a modest turnover of $62,000/m^3 of reactor volume.

The new production capacity that has been installed in Asia since the mid-1990s has allowed the fine chemical companies there to capture a share of more than 50% of the global merchant market for API-for-generics, a segment of the pharmaceutical industry, where demand has developed favorably in recent years (see Section 10.2).

10.2 DEMAND

The demand for fine chemicals depends mainly on the requirements of the largest customer base, the pharmaceutical industry. Its needs for custom manufacturing of pharmaceutical fine chemicals (PFCs) for proprietary drugs on the one hand, and for APIs-for-generics on the other hand, have developed in opposite directions. The innovator pharma industry's product portfolios are thinning because fewer new drugs are entering and more products are exiting it. In contrast, the generic companies are profiting from the trend to replace proprietary by nonproprietary drugs and the afflux of drugs that have lost patent protection.

Custom manufacturing, aka exclusives (see also Section 12.2.1), is driven primarily by the frequency of new drug launches and the industry's "make or buy" attitude; the latter is strongly influenced by the sourcing, resp. outsourcing policy in general and the available in-house production capacity in particular.

The generally used yardstick for new drug launches is the statistic of new drug approvals (NDAs) in the United States, published regularly by the FDA. As other agencies, such as EMEA (European Medicines Agency), are generally following the decisions of the FDA, they also are a good indicator of the frequency of global launches. The FDA's approvals of new drugs based on small molecules, the so-called new chemical entities (NCEs), has plunged from an all-time high of 51 in 1997 to an all-time low of 15 in 2005 (see Fig. 10.2). The big molecules, respectively new biological entities (NBEs), fared better. They plummeted from 8 in 1997 to 3 in 2005, but had a rebound in 2009. Obviously, the number of new APIs launched in any given year does not exactly correlate with the new sales generated. The individual products differ both in sales volumes and unit prices. Nevertheless, it is a valid indicator for the trend of the demand. The magnitude of the problem is illustrated by an

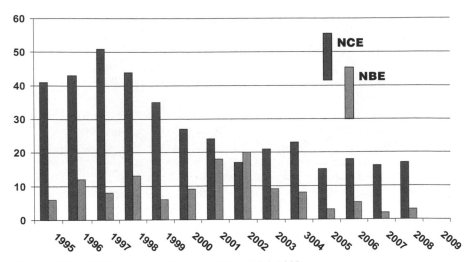

Figure 10.2 FDA approvals for new APIs, 1995–2008.
Source: FDA/Center for Drug Evaluation and Research (CDER).

evaluation done by The Boston Consulting Company (BCG). BCG found that a big pharma company must launch at least three new drugs per year, if it wants to maintain a growth rate in the low double-digit figure. Possible root causes for the slump in pharma R&D productivity are discussed in Section 15.2. Within exclusives, also known as "PFCs for proprietary drugs," the market for custom manufacturing has also suffered until recently from the reticence of the pharmaceutical industry, particularly "big pharma," to outsource the chemical manufacturing of their drug substances. Rather than being guided by a clear long-term strategy, their "make or buy" decisions are based on opportunistic short-term considerations.

An example in point is the contract manufacturing deal between Lonza and Bristol Myers-Squibb for Orencia (abatacept), a new drug for treatment of rheumatoid arthritis:

> In order to meet expected long-term demand for the drug, BMS has enlisted Switzerland's Lonza to share the manufacturing until the company builds a second manufacturing facility to cope with the output on its own.
>
> —*in-Pharma Technologist* (February 19, 2006).

A summary of the pros and cons for outsourcing from the pharma industry's perspective is given in Table 10.2. As extended studies at the Stern Business School of the New York City University have shown, financial considerations clearly favor the "buy" option [1, 2]: Two important findings of the studies were: "Focusing resources on chemical R&D instead of production can help

TABLE 10.2   Pros and Cons for Outsourcing API Manufacture

Pros	Cons
• **Concentrate on core activities (innovation and marketing)** • **Deploy your financial resources for more profitable investments** • Benefit from fine chemical industry's know-how and expertise • Eliminate long lead times to build and validate a manufacturing facility • Free capacity for new product introductions • Avoid risks of using hazardous chemistry	• Loss of tax benefits resulting from production of APIs in tax havens • Dissemination of Intellectual Property • Loss of know-how • Job losses • Underutilization of in-house production capacity • Overdependence on suppliers: "if we want 100% supply security, we have to do it in-house"

to increase returns" and "Return on physical assets in the chemical industry is roughly equal to the cost of capital, while the return on R&D substantially exceeds the cost of capital." Contrary to other big industries, both traditional ones, such as automotive, and modern, such as electronic, the pharma industry thus far was not under such a cost pressure to prioritize financial considerations. Actually, the recent slowdown in growth has given more emphasis on "avoiding job losses" and "filling underutilized capacity."

The latter has been created both by the abovementioned slump in new drug approvals and the megamergers in the industry. In-house manufacturing usually gets the right of first refusal for the manufacture of APIs. For new production, outsourcing is not even considered; old products that have been sourced from third parties are taken back in-house. Overall, the clash between an increased number of players and a boost of global production capacity on the one hand, and an overall sluggish demand on the other hand, has created a highly competitive environment.

This situation persisted until 2005–2007, when all big pharma companies finally started facing the fact that they needed to be more efficient and embarked on restructuring programs, of which outsourcing is an important part. The main beneficiaries of the additional business generated are the Asian fine chemical companies. The perspectives will be discussed in Section 16.1, Restructuring and Outsourcing.

Nonexclusives (see Section 12.2.2), especially *API-for-generics*, constitute the second most important outlet for fine chemicals after custom manufacturing. Because of new distribution systems, such as the health management (maintenance) organizations (HMOs) and government measures to contain costs of medication, global sales of generics are increasing rapidly. They are expected to reach $100 billion in 2010, or 12% of the total pharma market. On an API volume/volume basis, their market share will be close to 30%, due to the much lower unit prices for generics.

There are three main reasons for the booming generics market: (1) almost ubiquitous government pressure on reducing health-care costs—and therefore incentives for substituting originator drugs with generics; (2) the increasing consumption of Western medicine in developing countries; and (3) the large number of drugs that will lose patent protection over the years to come. In both the United States and Germany, more than 50% of physician prescriptions already are for generics—and the other countries are catching up rapidly. More than 60 "top 200 drugs" alone, representing aggregated sales of $140 billion, are expected to fall into the public domain by 2015. In terms of forthcoming API sales, this figure represents a total business potential of $2.0–2.5 billion (see Section 12.2.2).

As shown in Table 9.3, the merchant market for API-for-generics is estimated at $6 billion for 2010. The large domestic markets in China and India are not included in these figures. Asian companies in particular vie for the forthcoming generics business. They have the triple advantage of their low-cost basis, their big home markets, and previous manufacturing experience in producing for their domestic and other nonregulated markets. This is even more so in the case of APIs for the established generics. Apart from the rapid expansion of the Indian fine chemical manufacturing capacity (see Section 12.2.2), drug master file (DMF) submissions by Indian companies have also risen rapidly (see Section 14.1).

> DMFs are generic dossiers filed with the FDA in order to allow the API to appear in marketed drugs. Thus, an API manufacturer files just one application for a product, which can then be used to support approval of any generic based on that API.

The dichotomy in the development of offer and demand also has a direct impact on the development of the fine chemical industry in a global context. Whereas Western companies have suffered from the reduced demand for custom manufacturing services, Asian companies have benefited from the generics boom. The disparate turnover development of selected Indian and European fine chemical companies is depicted in Figure 10.1.

1. Fine Chemical Plant, F.I.S., Alte di Montecchio Maggiore, Italy

2. API plant, Bayer Schering Pharma, Bergkamen, Germany

3. PFC plant, Hikal, Bangalore, India

4. R&D Centre, Acoris, Pune, India

5. Kilogram Laboratory, Alphora Research, Mississauga, Canada

6. HPCL column, peptide plant, Lonza, Braine, Belgium

7. Biocatalysis pilot plant, Degussa-Evonik, Hanau/Wolfgang, Germany

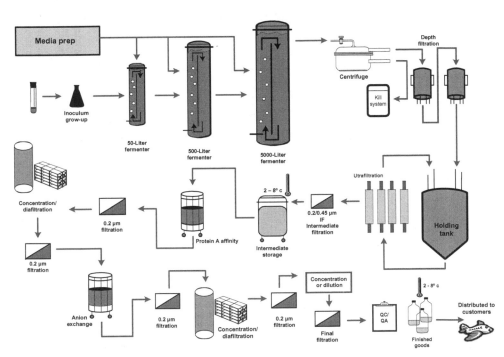

8. 5'000-L Process for Protein Production from Mammalian Cells

9. Airlift fermentor, fermentation pilot plant, Lonza, Slough, UK

10. Biologics plant, Lonza, Portsmouth NH, USA

11. PFC launch plant, Lonza, Visp, Switzerland

12. API plant, Boehringer Ingelheim, Ingelheim, Germany

13. HPAI plant, Helsinn, Biasca, Switzerland

14. SMB plant, Saltigo, Leverkusen, Germany

15. Kilogram Laboratory, Boehringer Ingelheim, Ingelheim, Germany

16. Preparative Laboratory, Acoris, Puna, India

17. PFC plant, Binhai New Area, Tianjin, China

18. Pilot Plant, Acoris, Pune, India

19. PFC plant, Hikal, Bangalore, India

20. 20000 Liter bioreactor, Lonza, Portsmouth NH, USA

CHAPTER 11

Customer Base

11.1 PHARMACEUTICAL INDUSTRY

The pharmaceutical industry constitutes the most important customer base for the fine chemical industry. It also has a track record of above-average growth. Global sales tripled from $52 billion in 1980 to $167 billion in 1990, corresponding to a compound annual growth rate (CAGR) of 12% per year. The growth slowed down in the decade 1991–2000 to—a still impressive—7.4% per year, arriving at $347 billion in the year 2000. The first half of the 2001–2010 decade still showed a strong growth, namely a CAGR of 9.5%. Because of an increasing generics penetration and decreasing number of new product launches, it slowed down to approximately 5% in the most recent 2006–2010 period. The total (i.e., captive plus merchant) value of pharmaceutical fine chemicals (PFCs) more or less follows the same trajectory, albeit at a 20 times lower level. As modern drugs are more active, dosages decrease and, consequently, the development of demand in terms of volume lags behind.

The pharma industry distinguishes itself from all other industries by its high R&D expenditure of 15–20% of sales. Search for a new medicine is complex, is long-drawn, and is fraught with a high degree of failure at any stage of the 10 to 14-year process (see Fig. 11.1). In order to test a new drug candidate's safety, efficacy, and dosage, clinical trials with thousands of volunteers have to be carried out, either with the target drug or a placebo. Around 5000–10,000 new chemical entities (NCEs) have to be synthesized in order to find out one day that one of them can be commercialized. The total costs for developing one new drug is $1.4 billion and therefore can only be afforded by "big pharma" companies (see below). For efforts to improve the R&D productivity, see Section 16.2.

The global pharma industry is conveniently subdivided into a three-tier structure: "big pharma," "medium pharma," and "small" or "virtual pharma." Whereas "big pharma" is active in R&D, manufacturing, and marketing, marketing prevails in "medium pharma." "Small pharma" companies focus on R&D (see Table 11.1). U.S. "big pharma" companies have outpaced their once

Fine Chemicals: The Industry and the Business, Second Edition, by Peter Pollak
Copyright © 2011 John Wiley & Sons, Inc.

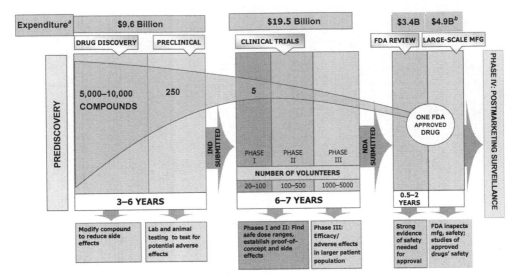

Figure 11.1 The drug discovery process.
[a] Amount spent in 2005.
[b] This figure includes Phase IV testing only.
Source: Enam Research, Pharma, India, 2010.

TABLE 11.1 Structure of the Pharmaceutical Industry

Type	Number	Characteristics	Examples
Big pharma	<20	Global companies with sales >$5 billion, many blockbusters, large in-house capabilities in R&D, manufacturing, and marketing.	See Table 11.2
Medium pharma	50–100	Sales $0.5–$5 billion, regional reach, <300 m^3 reactor volume, limited R&D capabilities, mainly in-licensed drugs and generics.	Barr, Forest, Shire, USA; Altana, Stada, Germany; Eisai, Japan; CSPC Pharma, China; Cipla, Ranbaxy, India
Virtual pharma	4–5000	Venture capital funded, 1–2 dev. drugs, mainly biopharmaceuticals, small R&D, no production, no M&S[a] organization.	Cell Therapeutics, Trimeris, Vertex, USA; Alizyme, UK; Morphosys, Germany; Basilea, Speedel, Switzerland

[a] Marketing and sales.

dominant European counterparts because of their marketing supremacy rather than a higher rate of innovation." For instance, each of them employs a "field force" of up to 20,000 (!) sales representatives, who visit physicians and hospitals at a heavy cadence. The total number of sales representatives amounts to about 100,000 equaling the total number of physicians. The aggressive marketing methods, which also comprise heavy "direct-to-consumer" advertising, have impaired the public image. The industry is criticized for promoting expensive new drugs that do not give additional therapeutic benefits over existing ones, covering up negative side effects, suppressing clinical studies with unfavorable results, bribing physicians, disease-mongering (i.e., inventing diseases), or using legislative loopholes to extend the patent duration of proprietary drugs [3].

"Medium pharma" are companies with annual sales in the $0.5–$5 billion bracket, with primarily regional reach. Most of the generics companies belong to this category. Consequently, the product portfolios consist primarily of generics. Occasionally, however, they succeed in launching a blockbuster drug, such as those of Altana Pharma (Germany) with Pantozol (pantoprazol), Lundbeck (Denmark) with Celexa (citalopram), and Sankyo (Japan) with Pravachol (pravastatin). They also draw advantage from the fact that it is no longer possible for "big pharma" companies to effectively pursue products that would be considered of minimal-to-limited turnover potential. This creates a reservoir of opportunities for the midsize drug companies, which exploit the potential of drugs in the $20–$200 million annual sales range. Many generics fall into this category.

The distinction between some generics companies and traditional, innovative pharmaceutical companies are blurred. On the one hand, the large generics companies move upstream, invest in R&D, and attempt to develop innovator drugs of their own as well. Thus, 30% of sales of the world's number one generics company, Teva, derive from innovative drugs. On the other hand, ethical pharma companies use generics, either in combination drugs (in the "sartan" category antihypertensives, e.g., both Merck's Hyzaar and Novartis' Diovan HCT are combination drugs between a proprietary active pharmaceutical ingredient [API] [losartan, resp. valsartan] and the generic diuretic hydrochlorothiazide), or in their consumer health over-the-counter (OTC) drug businesses. The many generic brands of N-acetyl-p-aminophenol and acetylsalicylic acid (Acuprin, Amigesic, Anacin, Anaflex, Arthritis pain Ascription, etc.) illustrate this. In a bid to stave off generic competition, some pharma companies also have introduced the concept of "authorized" generics. Here the brand (innovator) company "authorizes" a generic company to market the brand product as a generic under a different label. This ensures that new generics, enjoying a 180-day exclusivity period, will not face immediate competition, and opens up a new revenue stream for the innovator company. Yet another strategy adopted by pharmaceutical companies is to set up generic subsidiaries. Novartis has made a full-blown commitment to generics with its Sandoz subsidiary. Similarly, Abbott and Pfizer have created "Established Product" Divisions, and Sanofi-Aventis is engaged in generics through its Winthrop unit. Last but not least, in order to take advantage of the rapidly developing demand for drugs in the pharmemerging countries, Western pharma companies enter into alliances with local generic companies. AstraZeneca's licensing-and-supply deal with India's Torrent Pharmaceuticals, GlaxoSmithKline's marketing deal with Dr. Reddy, Pfizer's collaboration with Aurobindo and Strides, and Abbott's acquisition of Piramal's Healthcare Solution business follow the same blueprint.

Some big and medium pharma companies are hybrids in the sense that they also offer PFCs and custom manufacturing services and therefore compete with their PFC suppliers.

> Examples are Bayer Schering Pharma, Germany (through the acquisition of Schering AG by Bayer Health Care); Boehringer-Ingelheim, Germany (Industrial Chemicals division); Merck & Co., USA (through the acquisition of Schering-Plough, which in turn had acquired Organon Biosciences); Merck KGaA, Germany; Pfizer's Centre Source business (through the acquisition of Pharmacia Upjohn); and Sanofi-Aventis, France (Winthrop division, through the integration of Roussel-Uclaf).

The roots of the big and medium pharma companies go back either to drugstores or chemical companies: the former prevailing in the United States, while the latter in European enterprises. The lack of a chemical manufacturing history is also a reason why U.S. pharma companies *en principe* are more prone to outsourcing. Small pharma companies originate mostly from academia. Therefore, their R&D strategy is more focused on the elucidation of the biological roots of diseases rather than high-throughput screening (see Section 16.3). In order to attract investors, they are also referred to as "biopharmaceutical companies." The term is misleading, if used by firms developing small molecules. The attractiveness of big, medium, and small pharma as potential customers is described in Section 12.3, Target Markets: Geographic Regions and Customer Categories.

The ranking of the top 10 innovator and generic drug companies is shown in Table 11.2. The top 10 innovator pharma companies achieved aggregated sales of $300.3 billion in their core activities, or more than 40% of global sales of innovator drugs. This high degree of consolidation is due to the intensive mergers and acquisitions (M&A) activity that has taken place. There are five U.S. and European companies each among the top 10, whereby the European firms have a share of 56%, against 44% of their U.S. counterparts. However, these market shares do not take into account the fact that Swiss-based Roche has leapfrogged in the world's number 2 position due to the takeover of U.S. Genentech. ... Other major acquisitions in the recent past have been Merck taking over Schering-Plough and Pfizer absorbing Wyeth.

Teva and Sandoz and are by far the largest generics companies. They differ from their competitors not only in sales revenues but also because they are strongly backward integrated, are located outside the United States, and are not pure generics player companies. They also vie for the promising biosimilars market (see Section 17.2). In terms of both overall sales, profit growth and new product launches, the Israel-based Teva has been the frontrunner in recent years. Sales increased 2.5-fold from $5.3 billion in 2005 to $13.9 billion in 2009.

TABLE 11.2 Top 10 Innovator and Generic Pharma Companies

Innovator Pharma Companies			Generic Pharma Companies (or Divisions)	
Company	Sales 2009 ($ billion)		Company	Sales 2009 ($ billion)
	Total	Share of Ethical		
Pfizer, USA	50.0	45.4[a]	Teva,[b] Israel	16.2
Roche, Switzerland	47.4	37.7	Sandoz, Switzerland	7.5
GlaxoSmithKline, UK	44.2	37.0	Mylan, USA	5.1
Sanofi Aventis, France	42.6	35.1	Watson, USA	3.2
AstraZeneca, UK	32.8	32.3	Hospira, USA	2.7
Novartis, Switzerland	44.3	28.5	Ratiopharm,[b] Germany	2.5
Merck & Co., USA	27.4	25.3	Stada, Germany	2.2
Johnson & Johnson, USA	61.9	22.5	Sanofi, France	2.2
Eli Lilly, USA	21.8	19.9	Actavis, Iceland	1.7
Abbott, USA	30.8	16.5	Ranbaxy, India	1.5
Total Top 10[c]		**300**	**Total Top 10**	**45**

[a] Biopharmaceuticals, includes established products = generics and OTC.
[b] After the acquisition of Ratiopharm.
[c] **11–20:** Takeda, Japan (17.0/16.3); Bayer Health, Germany (23.0/15.1); Amgen, USA (14.6/14.6); Boehringer-Ingelheim, Germany (18.3/14.6), Bristol-Myers Squibb, USA (18.8/14.2); Daiichi-Sankyo, Japan (13.1[E]/8.5); Astellas, Japan (10.8/8[E]), Novo Nordisk, Denmark (9.9/8.2); Otsuka, Japan (12/7.9); Eisai, Japan (8.8/8.5).
$\Sigma_{11-20} \approx$ \$117 billion (Firma #20 fehlt noch!); $\Sigma\Sigma_{1-20} \approx$ \$417 billion.
Note: Activities outside ethical pharma comprise animal health, consumer health, diagnostics, generics, medical devices, nutritionals, OTC drugs.
Source: Company Annual Reports 2009.

The already strong position on the U.S. market was further extended by the acquisition of Ivax and Barr. "First to file" generics for proprietary drugs with total sales of \$25 billion are in the pipeline. Contrary to most of its competitors, Teva also has an R&D program for developing its own originator drugs and considerable in-house chemical manufacturing capabilities. Teva has 21 API manufacturing sites worldwide. Teva's Copaxone (glatiramer), the top-selling proprietary drug for treating multiple sclerosis on the U.S. market, achieved sales of \$2.8 billion in 2005. In the future, the company wants to maintain a 70/30 ratio between generic and originator drugs, respectively. The number 2 company is Sandoz. The name goes back to the Swiss chemical/pharmaceutical company Sandoz, which merged in 1996 with Ciba-Geigy to form Novartis. Novartis' generics are sold under the trademark Sandoz. Sales have increased at a rate of 12% p.a. from \$4.7 billion in 2005 to \$7.5 billion in 2009. Acquisitions of competitors Hexal and EBEWE Pharma contributed heavily. Forty percent of the revenues are already generated in emerging countries. The product portfolio of penicillins is backed up by captive production in plants in Kundl

(Austria) and in Torre Annunziata (Italy). Like Teva, Sandoz introduces years with a double-digit number (25 in 2009) of new generic versions of originator drugs per year. The other players all have strong positions in selected markets. As an indication of the things to come, Ranbaxy, an Indian company, ranks among the top 10 for the first time.

Pharmaceuticals containing more than 2000 different active ingredients are in commerce today; a sizable number of them are sourced from the fine chemical industry. They can be classified according to different criteria. In order of relevance to the fine chemical industry, they are described in the following paragraphs:

1. *Sales Volume.* All stakeholders in the pharmaceutical industry, the fine chemical industry included, have a keen interest in the top-selling or "blockbuster" drugs. The term "blockbusters" refers to pharmaceuticals with worldwide annual sales in excess of $1 billion. Their number has increased steadily, from 27 in 1999 to 51 in 2001, 76 in 2003, 103 in 2005, and ... then started leveling off. Key data of the top 10 blockbuster drugs are listed in Table 11.3, and their structural formulas are presented

TABLE 11.3   Top 10 Proprietary Drugs

Brand	API	Company	Sales 2009 ($ billion)
Lipitor	Atorvastatin (I)	Pfizer	12.5
Plavix	Clopidogrel (II)	Bristol-Myers Squibb Sanofi-Aventis	10.0
Enbrel	Etanecerpt HMW	Amgen, Pfizer, Takeda	8.0
Advair/Seretide	Salmeterol + (IIIa) Fluticasone (IIIb)	GlaxoSmithKline	7.8
Remicade	Infliximab HMW	Johnson & Johnson, Merck	6.9
Avastin	Bevacizumab HMW	Roche	6.0
Diovan & Co-Diovan	Valsartan (IV)	Novartis	6.0
Mabthera/Rituxan	Rituximab HMW	Roche	5.9
Abilifly	Aripiprazol (V)	Bristol-Myers Squibb Sanofi-Aventis	5.6
Humira	Adalimubab HMW	Abbott	5.5
Total Top 10[a]			**74.2**

[a] 11–20: (Figures in $ billion) Herceptin (trastuzumab), Roche, $5.1; Nexium (esomeprazol), AstraZeneca, 5.0; Zyprexa (olanzapine), Eli Lilly, 4.9; Seroquel (quetiapine), AZ, Astrella,4.9; Crestor (rosuvastatin), AZ, 4.7; Singulair (montelukast), Merck, 4.7; Neulasta/Neupogen (PEGfilgrastin/filgastrin), Amgen, 4.6; Effexor (venlafaxin), Pfizer, 4.3; Lantus (insulin glargine), Sanofi-Aventis, 4.3; Lovenox (enoxaparin), Sanofi-Aventis, 4.2.
Σ_{11-20} = $46.7 billion; $\Sigma\Sigma_{1-20}$ = $120.9 billion.
Source: Knol, Company Annual Reports 2009.

TABLE 11.4   Top Five Small-Molecule Proprietary Drugs/Structural Formula

(I) Atorvastatin

(IIIb) Fluticasone proprionate

(II) Clopidogrel

(IV) Valsartan

(IIIa) Salmeterol

(V) Aripiprazole

Note: Only the structural formula of the LMW APIs are shown.
Source: The Merck Index.

in Table 11.4. Five of the top 10 and 8 of the top 20 best-selling drugs are high-molecular-weight (HMW), aka biopharmaceuticals. However, with an average molecular weight exceeding 400, the "small" low-molecular-weight (LMW) molecules also exhibit quite complex structures. On average, they show three cyclic moieties (see also paragraph 5, below). Within 10 out of the top 12, at least one is *N*-heterocyclic. The largest-selling nonproprietary drugs are paracetamol, omeprazole, ethinylestradiol, amoxicillin, pyridoxine, and ascorbic acid.

2. *Position in the Drug Life Cycle*. A drug's position in the life cycle impacts distinctly the interface with a fine chemical company. The consequences for the selection of target products are discussed in detail in Section 12.2.

3. *Patent Status*. Drugs produced under valid product and manufacturing patents are referred to as *patented*, *proprietary*, *innovator*, or *ethical*

drugs. They account for the largest part of global drug sales. The leading pharmaceutical companies are active primarily in this category. It has lost part of its glory, as fewer new products arrive and more old products drop out. Drugs whose patents have expired are called *off-patent*, *nonproprietary drugs*, or *generics*. A generic drug (short: *generic*) is a drug that is bioequivalent to an originator drug but is sold for a lower price. Generics must contain the same active ingredient and the same strength as the originator drug. Generics can be legally produced and sold, if the patent of the innovator drug has expired or in countries that do not adhere to international patent legislation. They represent the largest number of active substances. They have smaller profit margins than do the proprietary drugs. The attractiveness of proprietary versus nonproprietary drugs for the fine chemical industry is discussed in Section 12.2.

4. *Sales Channels.* Prescription drugs can only be purchased by the public at a pharmacy with a physician's prescription. OTC drugs are available at convenience stores. Most OTC drugs are generics.

5. *Molecular Structure.* One distinguishes primarily between low-molecular-weight (LMW), or small, and high-molecular-weight (HMW), or big molecules. Most fine chemical companies are active in small molecules, which are produced by traditional chemical synthesis in conventional multipurpose plants, biocatalysis, microbial fermentation, or a combination of the three technologies (see Chapter 4). Within small molecules, *N*-heterocycles represent the most important class of pharmaceutical fine chemicals (PFCs) (see Section 3.1). An example is Boehringer-Ingelheim's Pradaxa (dabigran etexilate). It is the first new anticoagulant after the discovery of Warfarin in 1954. FDA approval is expected in 2010. The molecule has three cycloaromatic moieties, two of which, benzimidazole and pyridine, are *N*-heterocycles. The structural formula of Pradaxa is:

β-Alanin,N-[[2-[[[4-[[[(hexyloxy)carbonyl]amino]iminomethyl]phenyl] amino] methyl]-1-methyl-1*H*-benzimidazol-5-yl]carbonyl]-N-2-pyridinyl-, ethyl ester, methane sulfonate.

Big molecules, or biopharmaceuticals, have grown more rapidly than small ones. They have captured a of 10–15% share of the global pharmaceutical market (see Section 15.2). Only a few fine chemical companies possess the assets and the know-how for their production, primarily by mammalian cell culture technologies (see Section 4.2). At the interface between small and big molecules, the "tides" (i.e., nucleotides and peptides) have gained attention as pharmacologically active substances lately. Smaller peptides, composed of up to 30–40 amino acids, can be manufactured by conventional chemical protecting, coupling, and/or deprotecting methods. Larger ones, such as calcitonin and epoetin alfa, are produced by microbial biotechnology. "Bio fine chemicals" made by the most modern biotechnological process, the mammalian cell culture, have played an increasingly important role within the pharma market since the mid-1990s. First-generation products were recombinant human growth hormone (rhGH) and insulin (rhinsulin).

6. *Therapeutic Categories.* This classification is important for the pharmaceutical industry, as many companies specialize in selected therapeutic classes. According to IMS Health statistics, cholesterol and triglyceride regulators have the largest market share, followed by antiulcerants (aka oncology drugs), antidepressants and mood stabilizers, antipsychotics, and angiotensin-II-antagonists (aka antihypertensives). This classification is less relevant to the fine chemical industry. It should only be noted that the ratio between small and big molecules varies according to the individual class. For instance, small molecules prevail in antihypertensives; large ones, in oncology drugs.

11.2 AGROCHEMICAL INDUSTRY

Agrochemical companies are the second largest users of fine chemicals. Most products have a "pharmaceutical heritage." The process for the development of new agrochemicals is basically similar to that for pharmaceuticals. Apart from pesticides (mainly herbicides, insecticides, and fungicides), the leading companies are also involved in genetically modified (GM) crops, plants, and seeds. As a consequence of an intensive M&A activity over the past 10–20 years, the industry now is more consolidated than the pharmaceutical industry. The top 10 companies had a share of almost 95% of the total 2,000,000 tons/$44 billion pesticide output in 2008 (see Tables 11.5 and 11.6). Also, the locations of the top companies differ. Swiss and German companies hold the positions 1, 2, and 4, followed by Americans in the ranks 3, 5, and 6.

Generics play a bigger role in the agro than in the pharma industry. They represent some 70% of the global market. In further contrast to the pharma industry, the leading agrochemical companies themselves are heavily involved in generics. All top 10 agrochemical companies, except BASF, derive more than 50% of total sales revenues from generics. There are many agrochemical

TABLE 11.5 Top 10 Agrochemical Companies (or Divisions)

Company		Sales 2008 ($ billion)	Seeds/Biotech
1	Syngenta, Switzerland	9.23	++
2	Bayer CropScience, Germany	8.68	+
3	Monsanto, USA	5.33	+++
4	BASF, Germany	5.02	++
5	Dow Agrosciences, USA	4.06	+
6	DuPont, USA	2.64	+++[a]
7	Makhteshim-Agan, Israel	2.32	++
8	Nufarm, Australia	2.12	+
9	Sumitomo Chemical, Japan	1.36	+
10	Arysta LifeScience, Japan	1.23	+
	Total Top 10[b]	**$42 billion**	

[a] DuPont's Pioneer is the world's leading seed brand.
[b] **11–20**: (Figures in $ billion) FMC, $1.1; Cheminova, 0.96; United Phosphorous, 0.95; Ishihara (ISK), 0.48; Sankyo Agro, 0.40; Chemtura (Uniroyal), 0.39; Sipcam-Oxon, 0.39; Nippon Soda, 0.37; Nihon Nohyaku, 0.37; Kumiai, 0.33.
Σ_{11-20} = $5.7 billion; $\Sigma\Sigma_{1-20}$ = $47.5 billion.
Note: Non-crop sales (industrial, turf, home and garden, etc.) are not included.
Source: Cropnosis Ltd—Agranova.

TABLE 11.6 Top 10 Agrochemicals/Key Data

Brand	Active Ingredient	Company	Application	Sales 2008	
				$ billion[a]	MT
Round-up	Glyphosate (**I**)	Monsanto	Herbicide	8.30	620,000
Admire, Gaucho	Imidacloprid (**II**)	Bayer CropScience.	Insecticide	1.28	5450
Heritage	Azoxystrobin (**III**)	Syngenta	Fungicide	1.16	7000
F 500	Pyraclostrobin (**IV**)	BASF	Herbicide	1.10	7200
Flagship	Thiemethoxam (**V**)	Syngenta	Insecticide	0.73	1895
Callisto	Mesotrione (**VI**)	Syngenta	Herbicide	0.62	2040
Grammoxone	Paraquat-dichloride (**VII**)	Syngenta	Herbicide	0.60	26,000
Flint	Trifloxystrobin (**VIII**)	Bayer CropScience.	Fungicide	0.60	3405
Horizon, Folicur	Tebuconazole (**IX**)	Bayer CropScience.	Fungicide	0.55	2860
Regent MG, Frontline	Fipronil (**X**)	BASF	Insecticide	0.53	1375

[a] Ex-factory.
11–20: (Figures in $ million/MT) clothienidin (509/546); chlorpyrifos (482/34,945); chlorothalonil (475/48,559); lambda-cyanhalothrin (454/1085); 2,4-D (453/64,725); prothioconazole (417/1550); mesosulfuron-methyl (414/530); kresoxym-methyl (409/3450); acetochlor (400/39,000); glufosinate-ammonium (399/3990).
Source: Cropnosis Ltd—Agranova.

generics companies around the world. China National Chemical Corp., aka ChemChina Group, backed by the world's biggest buyout fund, Blackstone Group LP, agreed to pay $2.8 billion in cash for Nufarm Ltd. to form the world's largest supplier of generic farm chemicals in 2008. In terms of sales revenues, Mahkteshim Agan and Cheminova follow on the ranks 2 and 3. Apart from these multibillion dollar companies, there are hundreds of smaller firms with sales of less than $50 million per year, mainly in India and China.

The leading companies spend approximately 7.5% of their sales on R&D. A new crop protection product takes 9 to 10 years and $250–300 million to develop. In the period 2000–2009, 100 new LMW agrochemicals have been launched. However, only eight products achieved sales in excess of $100 million per year. Most successful were Bayer CropScience's prothioconazole and fluoxastrobin ($417 million and $223 million, respectively), followed by Syngenta's pinoxaden, DuPont's chlorantraniliprole, and Dow AgroSciences' aminopyralid. Since the 1990s, the R&D effort has focused mainly on next-generation agricultural biotechnology products. The global agro biotech market is $11 billion (2009), based on the sales price of biotech seeds plus technology fees. It comprises (1) GM crops with enhanced resistance to pests and environmental influences; (2) GM plants that contain beneficial ingredients such as oil or protein; and (3) GM plants that produce valuable complex products. Thanks to a major strategic thrust, and a liberal application policy, American companies hold a leading position in this area. At both Monsanto and DuPont's seed subsidiary, Pioneer Hi-Bred, GM seed businesses already account for more than 50% of total sales. To the detriment of the fine chemical industry, those two companies, therefore, have a limited share only of chemically produced pesticides. The third party requirement is further reduced by a high degree of backward integration. Whereas traditional agrochemicals are used mainly in cereals, maize, and rice, GM crops are most diffused in soybean, maize, and cotton.

In terms of applications, agrochemicals (synonyms: *pesticides*, *plant protection products*) are subdivided into herbicides, about 45–50%; insecticides, about 25–30%; and fungicides, about 20–25%. Nematicides, rodenticides, and fumigants account for the remaining 10%.

Examples within *herbicides* are the world's long-standing top-selling product, Monsanto's Round-up (glyphosate; formula (I) in Table 11.7). Round-up's patent expired in 2000, but Monsanto continued to dominate the market until 2008, when generic competition broke in at a faster rate than anticipated. The combination of modern application technology and cost leadership allowed Monsanto to fence off the Chinese competition until recently. Cost leadership was achieved by the switch to a much more economical process (see below) and the economy of size. Further leading pesticides are Syngenta's cyclohexadione-type mesotrione (VI) and good old paraquat dichloride (VII).

Within *insecticides*, the traditional organophosphates, such as malathion, and pyrethroïds, such as γ-cyhalothrin, cypermethrin, deltamethrin, flucythrinat, and prallethrin are being substituted by neonicotinoids, such as Bayer's

TABLE 11.7   Top 10 Agrochemicals/Structural Formula

(I) glyphosate

(II) imidacloprid

(III) azoxystrobin

(IV) pyraclostrobin

(V) thiamethoxam

(VI) mesotrione

(VII) paraquat-dichloride

(VIII) trifloxystrobin

(IX) tebuconazole

(X) fipronil

Source: Pesticide Manual.

imidacloprid (II) and Syngenta's thiamethoxam (V), and pyrazoles, such as BASF's fipronil (X). Aventis' Acetoprole, Cyazypyr (cyantraniliprole), launched 2008, is the latest addition to Du Pont's award-winning anthranilic diamide family of broad spectrum insecticides, the most important representative of which is Rynaxypyr.

Rynaxypyr (3-bromo-*N*-[4-chloro-2-methyl-6-[(methylamino)
carbonyl]phenyl]-1-(3-chloro-2-pyridinyl)-1H-pyrazole-5-carboxamide

Within *fungicides*, the strobilurins, a new class, are growing rapidly and already have captured more than 30% of the $10 billion global fungicide market. The first product, Syngenta's azoxystrobin (III), had been originally developed by Astra-Zeneca and launched in 1995. Also BASF F-500 Series, a.o. pyraclostrobin (IV) and kresoxim-methyl, Bayer CropScience, trifloxystrobin (VIII), and Monsanto are developing new compounds in this class. Older products are BASF's Opera, BayerCropScience's, Runner and Sportak, and DowAgro Science's Feniramol.

Combination pesticides are more and more frequently used, such as Monsanto's Genuity and SmartStax corn.

SmartStax combines the following herbicide-tolerant and insect-protection traits: YieldGard VT Rootworm/Roundup Ready 2 Technology & YieldGard VT PRO, Herculex I & Herculex RW and Liberty Link.

The largest market for pesticides is agriculture. Europe, North America, the Far East, and Latin America each contribute 20–25% of sales. Industrial applications (e.g., weed killers for railway tracks) and turf, home, and garden applications represent attractive niches for the agro industry but use relatively small quantities. Production volumes and application rates vary widely. Among the top 10 products, the volumes vary between over 600,000 MT for Monsanto's Round-up and 1400 tons for BASF's Regent MG, the median being about 5000 MT. The median application rate is in the 50–250 g/ha range, the extremes being 500–4000 g/ha for Round-up and 10–200 g/ha for Syngenta's Flagship. The top 10 agrochemicals in terms of sales in dollars are compiled in Tables 11.6 and 11.7. The list shows both modern products requiring small dosages and older ones with high dosages. The former products require much smaller application rates, but also have much higher unit prices. The most expensive agrochemicals in terms of unit prices are the modern insecticides γ-cyhalothrin (~$750/kg) and imidacloprid (~$500/kg); the cheapest ones are the old herbicides 2.4-D (~$6/kg) and Atrazin (~$5/kg). As shown in Table 11.7, the cheaper products are the simpler ones in terms of their chemical structure. Depending on the climatic conditions affecting crop yields and prices, the economic situation of the farmers, prices of agrochemicals are subject to wide fluctuations from year to year. Thus, the prices for Round-up varied between $5 and $20/gal over the past few years.

Active ingredients account for about 33% of the total turnover of agrochemicals, resulting in a production value of about $15 billion (see Table 9.2). The proportion of the active substance in the price of the formulated product, therefore, is much higher than in drugs. The main reason for this difference is the pressure on the price for agrochemicals. A farmer does not use a pesticide unless there is a distinct economic advantage in terms of better crop yields. Pharmaceuticals are not driven by the same cost/benefit imperative.

The active ingredients of older pesticides were rather simple molecules. Thus, one of the best-known classical herbicides, 2,4-D is just 2,4-Dichlorophenoxyacetic acid. As illustrated by the structural formulas of the top 10 agrochemicals, modern pesticides are more complex. Seven compounds contain one or more heterocyclic rings. Thus, similar to the situation in LMW PFCs, heterocyclic chemistry also plays an important role in agrochemicals. The average molecular weight of the top 10 is 330, not much less as compared with >400 for the top 10 pharmaceuticals.

Given the diversity of the chemical structures, a wide variety of chemical process technologies are also used for the manufacture of agro fine chemicals. An example is glyphosate. The pivotal intermediate is iminodiacetic acid. The conventional synthesis is a Strecker reaction from hydrogen cyanide, formaldehyde, and ammonia. It has been superseded by a catalytic oxidation of diethanolamine. The hydrogen atom in the imino moiety is then substituted by phosphonic acid, through chlorination with phosphorous trichloride and hydrolysis. In comparison to reagents used in PFC syntheses, hazardous chemicals, for example, sodium azide, halogens, methyl sulfide, phosgene, and phosphorous chlorides are more frequently used. Agrochemical companies sometimes outsource just these steps, which require specialized equipment, on toll conversion deals. With exception of pyrethroids, which are photostable modifications of naturally occurring pyrethrums, active ingredients of agrochemicals rarely are chiral.

For the future attractiveness of the agrochemical as partner for the fine chemical industry, see Section 18.1.

11.3 ANIMAL HEALTH INDUSTRY

Animal health is a segment of the life sciences industry at the interface of pharmaceuticals and agrochemicals. Global sales were $12.5 billion in 2003, $19.2 billion in 2008, and are estimated to reach $20.2 billion in 2011. The CAGR between 2003 and 2011^E is 7%. Of the top10 companies, eight are business units or spin-offs from pharmaceutical companies (see Table 11.8). The industry is rather concentrated, with the top 10 companies accounting for 76% of total sales. As they do in pharma, U.S. companies dominate in animal

TABLE 11.8   Top 10 Animal Health Companies or Divisions

Company		Sales 2008 ($ billion)		
		Total	./. Animal Health	./. Biologics
1	Intervet/Schering-Plough[b], USA	2.97	2.91	1.33
2	Pfizer, USA	2.82	2.77	0.59
3	Mérial[b], France	2.64	2.64	0.79
4	Bayer, Germany	1.40	1.28	0.02
5	Elanco, USA	1.11	1.09	0.08
6	Fort Dodge, USA	1.09	1.09	0
7	Novartis, Switzerland	1.09	1.04	0.56
8	Boehringer-Ingelheim, Germany	0.68	0.68	0.34
9	Virbac, France	0.65	0.62	0.07
10	CEVA, France	0.54	0.54	0.12
Total Top 10 Animal Health[a]			**14.7**	**3.9**

[a] 11–15: (Figures in $ million) CEVA, $340; Alpharma, $325; Phibro, $297; Vétoquinol, $244[E]; Monsanto, $225[E].
[b] Merger to Merial-Intervet is pending.
Source: Vetnosis Executive Guide 2009.

health products. Many of the veterinary products in the portfolios had originally been developed for human use or as pesticides. Thus, APIs of human
drugs are just reformulated to be ready for animals. Examples in the therapeutic class of anti-anxiety drugs are Eli Lilly's, Reconcile (fluoxetine), nick-
named "puppy Prozac," and Novartis' Clomicalm. Its active ingredient,
clomipramin, is also the API of Anafranil. Also, Pfizer's anti-obesity drug,
Slentrol (dirlotapide), had originally been developed for human use. The same
is the case for the active ingredients of new veterinary drugs for the treatment
of arthritis pain in dogs, namely, Mérial's Previcox (firocoxib), Novartis'
Deramaxx (deracoxib), and Pfizer's Rimadyl (carprofen). Previcox and
Deramaxx belong to the well-known class of COX-2 inhibitors. Rimadyl is a
nonsteroidal anti-inflammatory. The ratio LMW/HMW molecules is about
80/20.

The *farm animal*, or *livestock* segment, represents about 55% of the total.
Pigs constitute the biggest share in terms of numbers of animals; cattle, in
terms of veterinary drug consumption, followed by pigs, poultry, and sheep. In
terms of product categories, the animal health market is dominated by parasiticides (28%), mainly endectocides, such as Mérial's Ivermectin, Pfizer's
Doramectin, and Fort Dodge's Cydectin and antibiotics (17%), such as cephalosporins, penicillins, sulfonamides, and tetracyclines. They are widely used to
treat infections in both human and veterinary medicine. Classical examples
are ampicillin and amoxicillin, such as Intervet's Amfipen and Amoxypen,
respectively.

In the "pet" (respectively *companion animal*) segment, which comprises
cats, dogs, birds, some rodents, reptiles, and horses, and represents about

TABLE 11.9   Top 10 Veterinary Drugs 2009

	Brand	Active Ingredient	Company
1	Frontline	Monensin	Elanco
2	Advantage	Imidacloprid	Bayer
3	Rumensin	Monensin	Elanco
4	Heartgard	Ivermectin	Merial
5	Ingelvac Circoflex	Purified circovirus antigen[a]	Boehringer-Ingelheim
6	Revolution/Stronghold	Selamectin	Pfizer
7	Baytril	Enrofloxacin	Bayer
8	Tylan	Tylosin	Elanco
9	Rimadyl	Carprofen	Pfizer
10	Naxcel/Excenel/Excede	Ceftiofur	Pfizer

[a] Vaccine.

Source: Boehringer-Ingelheim.

40% of the total market, the association with human health is particularly prominent.

The third segment, *aquaculture*, represents about 5–10% of the market. Nowadays, about one-third of fish for food are produced by aquatic farming, where intensive care is mandatory.

The top 10 veterinary drugs are listed in Table 11.9. Their annual sales range between $100 million and $500 million.

The degree to which active ingredients are sourced from third parties varies considerably. It depends primarily on the affiliation of a given animal health business and the patent status of the product. Thus, animal health businesses that are divisions of pharma companies (6 out of the top 10) procure the active substances for their patented products primarily from the parent company. An example is the abovementioned Clomicalm. Once a product becomes non-proprietary, it is more likely purchased on the merchant market. This is, for example, the case for veterinary drugs containing ampicillin and amoxicillin. Small independent animal health companies generally outsource all their requirements for active substances. This also applies to the many small distributors of veterinary drugs, which do not have in-house chemical manufacturing capabilities.

Also, the proportion of the active ingredient in the price of the finished drug varies widely. It can be below 1% in low-dosage, rarely administered, patented injectable drugs. In contrast, it can be up to 25% in generic drugs administered as powders.

Whereas the market in the developed world is almost saturated, a rapid increase in demand is expected in developing countries. Of 1.5 billion farm animals, 1 million are located in tropical and subtropical countries. Also, more and more of the latter are raised by industrial methods and require regular medication. Likewise, an increasing number of affluent people in these countries can afford "lifestyle" medicines for their pets.

In terms of NCEs, the animal health market is not innovative. The patents of most active ingredients have expired. The regulatory hurdles for approval of new veterinary drugs are almost equally high as for human drugs. The average development time is 5 to 9 years. The pertinent regulatory bodies are the FDA for drugs and the Environmental Protection Agency (EPA) for parasiticides. The overall business risk is lower with an animal drug because testing on the target animals is done early in the clinical trial process. Relatively few active ingredients are made by traditional chemical synthesis. About 20% of all veterinary drugs are vaccines. New products are derived mainly from new pharmaceuticals that are formulated to the requirements of the animals. Thus, business opportunities for fine chemical companies arise when animal health divisions are spun off from their parent pharmaceutical companies and are therefore cut off from the supply of the active ingredients. This was the case when Mérial was created as a joint venture between Merck & Co.'s MSD AgVet division and Rhône Mérieux in 1997.

11.4   OTHER SPECIALTY CHEMICAL INDUSTRIES

Apart from the life sciences, specialty chemicals—and therefore also their active ingredients, commodities or fine chemicals, as the case may be—are used ubiquitously, in both industrial applications, such as biocides and corrosion inhibitors in cooling water towers, and consumer applications, such as personal care and household products. The embedded products extend from high-price/low-volume molecules used for liquid crystal displays, to large-volume/low-price amino acids used as feed additives.

Examples of applications in eight areas, ranging from adhesives to specialty polymers, are listed in Table 11.10.

The manufacturer base is very heterogeneous, too. It extends from small "garage-type" outfits, which prepare some simple household cleaner formulations, to the mega-size specialty companies, such as the "Four Soapers," Henkel, Germany; Kao Soap, Japan; Procter & Gamble, USA; and Unilever, The Netherlands, as well as the big food companies, Ajinomoto, Japan; Danone, France; Kraft Foods, USA; and Nestlé, Switzerland. The innovation process of the latter is directed mainly toward creation of new formulations. The development of NCEs is the exception rather than the rule. It is most likely to happen in application areas unrelated to human health (where NCEs are subject to very extensive testing). This is the case with electronic chemicals and specialty polymers.

The attractiveness of "nonlife science" specialties for the fine chemical industry varies considerably, both among the categories listed and within themselves. The determining elements are size of the market for the embedded active ingredients in terms of both volume and sophistication, the backward integration of the manufacturers, and the growth potential. Many specialty chemicals are "commoditized" and are based on low-priced commodities as active ingredients.

TABLE 11.10 Application Examples of Fine Chemicals outside Life Sciences

Application		Product		Manufacturer (Examples)
General Category	Specific	Brand Name	Chemical Name	
Adhesives	Rapid bonding	Super Glue	α-Cyano acrylates	Henkel (Loctite), Germany
	Structural adhesive	Araldite	Diglycidyl ether type	Huntsman
Biocides (industrial antimicrobials)	Industrial antimicrobial	Zonen-10	[5-Chloro-]2-methyl-isothiazoline-3-one	Chemicrea, Japan
	Disinfectant	Chlorohexidine digluconate	1,1'-Hexamethylenebis[5-(p-chlorophenyl)biguanide] digluconate	Evonik-Degussa, Germany; Medichem, Spain
Catalysts and enzymes	Water treatment	TBZ	Thiabendazole	Hikal, India
	Chiral synthesis	BINAP[a]	2,2'-Bis(diphenylphosphino)1,1'bin-1,1'-binaphthyl	Rhodia, France; Solvias, Switzerland
		Novozym 388	1,3-Specific lipase	Novozymes, Denmark
Dyestuffs and pigments	Cellulosic vat dye	Indanthren	6.15-Dihydro-5,14,18-anthrazinetetrone	Kiri Dyes & Chemicals, India
	Red automotive coating	Irgazin DPP Red	Diketopyrrolopyrrol	BASF, Germany
Electronic chemicals	Ink jet printer ink	PRO-JET	(see footnote)[b]	Fuji Photo Film (ex Avecia)
	Liquid crystals		e.g., Trifluoromethoxy-phenyl derivatives	Merck KGaA, Germany
	Etchants	Zyron 8020	Octafluorocyclobutane	Du Pont, USA
	OLED	Novaled Pin OLED	Various	Novaled, Germany; Plextronics, USA
Flavors and fragrances	Flavors	EuroVanillin Rhovalin	Vanilline	Borregaard, Norway Rhodia, France
	Fragrances	Jasmone Cis	cis-3-Methyl-3-(2-pentenyl)-2-cyclopenten-1-one	Givaudan, Switzerland

Food and feed additives	Dietary supplement	Carnipure tartrate	L-Carnitine L-tartrate	Lonza, Switzerland
	Vitamin feed additive	Biotin, vitamin H	Hexahydro-2-oxo-1H-thieno[3,4-d] imidazole-4-pentanoic acid	DSM, The Netherlands
	Amino acid feed additive	Biolys®	L-lysine	Evonik-Degussa, Germany
Specialty polymers	Aerospace composites	Primaset® BADCY, PT-resins	2,2-Bis(4-cyanatophenyl) propane and oligomers, Novolac-cyanates	Lonza, Switzerland
		Matrimid Kapton Avimid	Non-melting polyimides	Huntsman, USA DuPont, USA Mitsui, Japan
	Electronic composites	Primaset® BADCY, PT-Resins	2,2-Bis(4-cyanatophenyl) propane and oligomers, Novolac cyanates	Lonza, Switzerland
		BT-Resins	2,2,-Bis(4-cyanatophenyl) propane and oligomers	Mitsubishi, Japan
		Matrimid Kapton Avimid	Non-melting polyimides	Huntsman, USA DuPont, USA Mitsui, Japan
	Wire and cable	Teflon FEP Neopflon FEP	Fluorinated ethylene/propylene polymer	DuPont, USA Daikin, Japan
	Electronic coatings	BCB	Benzocyclobutene polymers	DOW, USA
		Ultrem, Siltrem	Polyimides	GE-Plastics, USA
	Precision parts, electric connectors	Vectra Zylon Twaron	Liquid crystalline polymers	DuPont, USA Toyobo, Japan Akzo-Nobel, The Netherlands

---

[a] Inventor: R. Noyori.

[b] Direct Blue 199 for cyan; combination of Direct Yellow 86 and 132 for yellow; combination of Acid Red 52 and Reactive Red 180 for magenta, Fod Black 2 for black.

11.4.1 Adhesives and Sealants

Adhesives and sealants are valued at approximately $60 billion (2009), of which specialty adhesives and sealants about $12 billion, are growing globally at a rate of about 3% per year. Lower volume, high-value applications such as electronics assembly and flat panel displays will experience the strongest growth. The leading players are Henkel, $10.2 billion; Sika, $2.9 billion; 3M, $2 billion; Bostik, $2 billion; H.B. Fuller, $1.4 billion; Mapei (Milan), $1.2 billion; Illinois Tool Works (Glenview, IL), $1.2 billion; RPM, $1.1 billion; and Dow's Rohm and Haas subsidiary, $0.7 billion (all numbers refer to 2007).

The applications extend from household, such as paper gluing, all the way to high-tech specialty products used for assembling electronic parts and automotive and aircraft construction. In the latter application, they are increasingly substituting for welding. Most of the products are chemically uncomplicated compounds, such as the phenolics, poly(vinyl acetate/alcohol), rubber cements, melamine-, phenol-, and urea-formaldehyde, and most epoxy resin types. Fine chemicals are used for "high-tech" adhesives, such as the α-cyanoacrylates, which are used for assembling electronic components and parts of modern civil and military aircraft, including rotor tips for high-speed helicopters.

11.4.2 Biocides

Biocides also cover a wide array of applications. Within the approximately $2.5 billion global market, the largest applications are for wood conversation and water treatment (drinking water, cooling towers, swimming pools) followed by paints and coatings. Applications include anti-mildews, antimicrobials, bactericides, disinfectants for hard (hospitals, schools, restaurants) and soft (skin) surfaces, preservatives for cosmetics, food, marine coatings and water-based paints, sanitizers, and slimicides. Preservatives are usually the highest-value biocides. It is often necessary to include a variety of preservatives to deal with mold, bacteria, fungus potential, and other contaminants. The inclusion rate is very small, less than 1%, but the value per kilogram can be high, especially in high-priced cosmetics.

In order of increasing carbon atom count, *organic* biocide active ingredients extend from C_1–C_3 petrochemicals through quaternary ammonium compounds, brominated aliphatics, phenolics, N-heterocycles, and N,S-heterocycles, all the way up to biguanide polymers used for swimming pool sanitation. Selected important biocides are listed in Table 11.11.

Examples of products based on *inorganic* active ingredients are the century-old "Eau de Javelle" used for removing stains from shirt collars and terracotta floors, and iodine for disinfecting small bruises.

Chemical companies have their own brands of biocides. For example, AkzoNobel proposes Arquad and Berol; Chemtura, the Bellacide; Evonik, the Quab; Dow, the Dowicide; DuPont, the RelyOn; Lonza, the Bardac and Barquat; and Rohm and Haas, the Kathon biocide product line. Rather than

TABLE 11.11 Biocides

Brand Name	Chemical Name	Manufacturer	Application
	C_0–C_{10}		
	Iodine	Nippoh	Disinfectant
Eau de Javelle	Sodium hypochlorite	Solvay	Disinfectant
RelyOn	Potassium peroxymonosulfate	DuPont	Disinfectant
Formaline	Formaldehyde	SABIC	Preservative
Ethyl alcohol	Ethanol	ADM	Disinfectant
Isopropyl alcohol	2-Propanol	Shell	Disinfectant
Protectol BN	2-Bromo-2-nitropropane-1.3-diol	BASF	Antibacterial
DBNPA	Dibromo nitrilo propionamide	Dow	Biocide
Zonen-10 Kathon	5-Chloro-2-methyl-4-isothiazoline-3-one	Chemicrea; Rohm&Haas	Biocide
Protectol DZ	3.5-Dimethyl-1.3.5-thiadiazinane-2-thione	BASF	Preservative
Cidex	Glutaraldehyde	Prestige Brands	Biocide
Dantogard	DMDMH (Dimethylol dimethyl hydantoin)	Lonza	Biocide
Chloraseptic	Sodium phenolate		Antiseptic
Ferulic acid	4-Hydroxy-3-methoxy-cinnamic acid	esp. Chinese	Preservative
	C_{11}–C_{20}		
Butylparaben	Butyl 4-hydroxy benzoate	Merck	Preservative
Dowicide A	o-Phenyl-phenol	Dow	Biocide
pHisohex	2,2′-Methylenebis (3.4.6-trichlorophenol); hexachlorophene	Sanofi-Aventis (formerly Givaudan)	Biocide
Trimethoprim	5-[(3.4.5-Trimethoxyphenyl) methyl]-2.4-pyrimidinediamine	Roche	Antibacterial
	C_{21}–C_{50}		
Bellacide	4-Chloro-2.6-dialkylamino-s-triazine	Chemtura	Biocide
Bardac 22	Didecyl dimethyl ammonium chloride	Lonza, Stephan	Biocide
Chlorhexamed Merfen	Chlorhexidine[a] digluconate (C_{34})	Degussa, GSK, Novartis	Disinfectant
Tylan	Tylosin (macrolide, C_{46})	Eli Lilly	Antibacterial
Polymers			
Povidone-I_2 Betadyne	Poly(vinylpyrrolidone)/iodine	Purdue Pharma[b]	Disinfectant
Vantocil	Poly(hexamethylenebiguanide)	Arch Biocides	Antimicrobial

[a] Chlorhexidine = *N,N″*-Bis(4-chlorophenyl)-3,12-diimino-2,4,11,13-tetraazatetradecanediimideamide.
[b] Manufacturers of the PVP segment are BASF and ISP.

competing on chemicals alone, the manufacturers are shifting toward enhanced service provision.

As with other specialty chemicals, which can impair human health and therefore need extensive and expensive testing prior to commercialization, R&D in biocides is concentrating on formulations and applications, rather than on NCEs. NCEs are developed for substitution of old products with environmental hazards, such as the copper/arsenic compounds used in timber and wood preservation.

11.4.3   Catalysts and Enzymes

The defining work that set the stage for modern catalytic technology was the industrial-scale production of ammonia from hydrogen and nitrogen over an iron-based catalyst via the Haber-Bosch process at BASF in Germany in 1913. This led to the production of synthetic fertilizer, which had an enormous effect on human population growth. It is estimated that 40% of the population would not exist without it. Since then, approximately 20% of the world economy depends on catalytic processes that generate >95% of all chemical products by volume and >80% of their added value. The total market for *catalysts and enzymes* amounts to $15 billion ($2010^E$). There are four main applications: "environment" (e.g., automotive catalysts), 31%; "polymers" (e.g., polyethylene and polypropylene), 24%; "petroleum processing" (e.g., cracking and reforming), 23%; and "chemicals," 22%. Catalysts have a share of 80%, enzymes 20%.

The leading international players in catalysts are Evonik-Degussa (Germany), BASF (Germany), Süd-Chemie (Germany), Johnson Matthey (UK), CRI/Criterion (part of the Shell group from the UK and The Netherlands), Haldor Topsoe (Denmark), WR Grace (USA), Nippon Shokubai (Japan), and Nikki Chemicals (Japan). Evonik-Degussa offers powder precious and base (Raney-type) metal, fixed bed, olefin polymerization, and homogeneous metathesis catalysts for the life science, fine chemical, bulk chemical, and polymer industries. BASF has products in the areas of powder precious metal, supported nickel, fixed bed, and automotive catalysts. Johnson Matthey has the same offerings as BASF with the addition of Raney-type nickel and homogeneous catalysts. WR Grace produces fixed bed, Raney base metal, and polyolefin catalysts, while Süd-Chemie, CRI/Criterion, Haldor Topsoe, and Nippon Shokubai focus mostly on fixed-bed catalysts for petrochemical applications.

Novozymes is the largest conventional enzyme company (sales 2009 approximately $1.5 billion, market share approximately 45%) serving these markets: detergents, 30%; technical enzymes, 30%; food enzymes, 23%; feed enzymes, 9%; microorganisms, 4%; and biopharma, 4%. Danisco, Denmark, and DSM, The Netherlands, follow on positions numbers 2 and 3.

Within fine chemicals, which are parts of the a.m. category "chemicals," catalysts and enzymes play an important role as enablers for the synthesis of chiral molecules. Ligands such as derivatives of 2,2'-Bis(diphenylphosphino)-1, 1'-binaphthyls (BINAPs) are used with precious metals to form homogeneous catalysts that provide very high activity and selectivity for the production APIs and other chemicals where selectivity (e.g., chemo-selectivity, regio-selectivity, and chirality) as well as purity are important. Solvias, Switzerland, is a contract research organization (CRO) specializing in the development and application of BINAPs and other ligands for homogeneous catalysts. The most renowned industrial-scale applications are the manufacture of the fragrance ingredient *l*-menthol by Takasago, Japan, and that of the herbicide (*S*)-metolachlor by Syngenta, Switzerland. Recent developments with *N*-heterocyclic carbene and alkylidene ligand containing ruthenium homogeneous catalysts (e.g., the CatMETium products of Evonik-Degussa) have made metathesis reactions a viable route for the production of chemicals. Cross metathesis is particularly interesting for oleochemicals; ring-closing metathesis has found use in pharmaceutical applications such as the synthesis of anticancer agents, and ring-opening metathesis polymerization is generating new polymeric materials.

Precious metal powder catalysts applied in the slurry phase have excellent heat dissipation properties for exothermic reactions. They are used in various hydrogenation reactions such as aromatic nitro group reduction for the production of toluenediamine and aniline used in polyurethanes. They also perform hydrogenolysis reactions commonly used in the deprotection strategies for pharmaceuticals such as ramipril, lisinopril, enalapril, fosinopril, and benazepril. Another of many applications for precious metal powder catalysts is the reductive alkylation of amines with carbonyl compounds. A known commercial application is the reductive alkylation of *p*-aminodiphenylamine with methyl isobutyl ketone over a sulfided platinum on activated carbon catalyst for the production of *N*-(1,3-dimethylbutyl)-*N'*-phenyl-*p*-phenylene diamine, which is used as an antiozonant in tires. The powder base metal catalysts used in the slurry phase can either be supported (e.g., Ni/SiO_2 and Ni/Al_2O_3) or made by a technique first invented by Murray Raney in the late 1920s. The nickel catalysts made by the Murray Raney technique are particularly useful for the production of oleochemicals, the hydrogenation of sugars to polyols (e.g., sorbitol and maltitol) and nitrile groups to amines where the production of fatty amines for surfactants and hexamethylenediamine for nylon are excellent examples. A recent development with Raney-type catalysts is the Centoprime technology of Evonik-Degussa that features steric ensemble (the number of contiguous Ni atoms per active site) control for the chemo-selective hydrogenation of nitriles to primary amines in the presence of other moieties.[1] The Centoprime technology is currently being used by DSM Nutritional Products for hydrogenating pynitrile to the desired Grewe diamine for the production of vitamin B1.

---

[1] D. J. Ostgard, R. Olindo, M. Berweiler, and S. Röder, *Catalysis Today*, vol. 121, 2007, pp. 106–114.

Fixed-bed catalysts are used in continuous processes running in the gas, liquid, or trickle phases (similar to an aerosol that completely covers the catalyst with a thin layer of reaction medium). The technology is suitable for large volume products (more than ~7000 MT per year) requiring oxidation, hydration, dehydration, ammoxidation, hydrogenation, hydrodesulfurization, hydro-denitrogenation, or isomerization steps in their manufacture. An example of a fixed-bed partial oxidation is the acetoxylation of ethylene in oxygen with a potassium acetate doped Pd + Au catalyst for the manufacture of vinyl acetate used to produce polyvinyl acetate. Fixed-bed oxidation technology is also used in the stepwise conversion of propylene to first acrolein and then acrylic acid over two different mixed-metal oxide catalysts in series. The economics of fixed-bed hydration make it ideal for the cost-effective production of synthetic ethanol from ethylene and water over phosphoric acid impregnated silica.

Enzymes are HMW organic molecules (see Section 4.2). Standard enzymes, such as acylases, amylases, β-lactamases, cellulases, lipases, nitrilases, and proteases, are extracted from natural products. Some typical industrial applications for enzymes are the production acrylamide from acrylonitrile with nitrile hydratase, the isomerization of glucose to fructose with glucose isomerase for "high-fructose corn syrup," the hydrolysis of sucrose to a glucose and fructose mixture with invertase, and the hydrolysis of starch to either glucose with amyloglucosidase or mostly maltose with β-amylase. Most of the sugar and starch applications are with immobilized enzymes for ease of use. Specialty enzymes, which are tailored for specific reactions, constitute a recent advance. They improve the volumetric productivity 100-fold and more. A frequently debated modern industrial application is the production of "bioethanol" from cellulosic feedstocks. For application examples in PFC synthesis, see Figures 4.1 and 4.2. Fine chemical companies, which have an important captive use of enzymes, are offering them to third parties.

A few examples are described here:

DSM, through the acquisition of Gist-Brocades, has a strong position in enzymes for penicillin manufacture.

Codexis, USA and Protéus, France are companies specializing in the development of tailored enzymes. With its "molecular breeding" technology, Codexis splits the protein chains of existing enzymes into small peptide fragments. These are then reassembled randomly. Some of the artefacts thus obtained exhibit much better performance. Codexis won a 2006 Presidential Green Chemistry award for the directed evolution of three biocatalysts to produce ethyl (*R*)-4-cyano-3-hydroxybutyrate, "hydroxynitrile," the key chiral building block for Pfizer's Lipitor (atorvastatin). The starting material is ethyl 4-chloroacetoacetate (see also Table 1.1).

Protéus' know-how in recombinant expression in micro-organisms and process development has enabled the company to create a comprehensive platform of services to address recombinant protein expression challenges.

Producers of specialized catalysts or enzymes are faced with the dilemma that the better performing and the more specific their products are, the smaller is the market requirement. A few grams can suffice for the production of multitons of a fine chemical. Therefore, they extend their offerings by developing the pertinent production processes. Sometimes, they even enter into cooperations with fine chemical companies to produce target molecules. In any case, the continued advances in this technology can only be supported by either value-added pricing for the catalyst or by additional revenue streams such as licensing agreements.

11.4.4 Dyestuffs and Pigments

The serendipic discovery of mauveïne, also called aniline *purple*, by William Perkin in 1856 gave birth not only to the dyestuff industry but also the organic chemical industry as a whole. The origins of the pharmaceutical and agrochemical industries, too, go back to dyestuff manufacture. The *dyestuff and pigments* industry nowadays generates annual sales of $9–$10 billion. The molecules extend from simple aniline derivatives to sophisticated color shifting pigments, costing as much as $3000/kg and used a.o. in bills to prevent counterfeiting.

Substantial changes have occurred in the industry since the early 1990s. In the United States, the production of dyestuffs has practically disappeared. Whereas DuPont sold its organic pigments business (the company still produces the inorganic pigment titanium dioxide) to CIBA back in 1984, Sun Chemical, a division of the Japanese Dainippon Ink Corporation, is the only producer left. Europe, which had held a dominant position for more than 100 years, has lost out against China, India, and South Korea. DyStar, Germany, with a headcount of 3500, had been the world's largest dyestuff manufacturer. The company goes back to a merger between the Bayer and Hoechst dyestuff divisions in 1995, with BASF joining 5 years later. The company reported sales of $1.2 billion in 2009. It was rescued from insolvency by Kiri Dyes & Chemicals, Ahmedabad, India, for $80 million in 2010. The firm's largest plant, located at Brunsbüttel, produces 20,000 tons. Clariant, Switzerland, produces both dyestuffs and pigments. The leader in pigments is BASF, Germany. It acquired CIBA Specialties, Switzerland, in 2008. The latter had sold its dyestuff division to Huntsman in early 2006, but kept its pigment business. The dyestuff production in the Far East countries dwarfs the European one in terms of production volume. China's production of dyestuffs and organic pigments totaled more than 1,000,000 tons in 2009. India has a production capacity of 150,000 MT. Whereas Europe focuses on the production of the high-end specialties, simpler dyestuffs and pigments and their intermediates are now sourced from the Far East countries, primarily China, India, and South Korea. Three cooperation models are used: (1) establishing its own factories (DyStar has three of them in China); (2) entering into joint ventures, such as the CIBA Specialties/Indian Dyestuff Industries joint venture (JV); or (3) sourcing from the hundreds of "garage-type" producers. China benefits not only from a low-cost advantage but also from the fact that the country prides itself on having the world's largest textile market, which is the main outlet for dyestuffs. Statistics indicate that it is likely to account for 40% of the global output of textiles in the medium term. The largest dyestuff company in the Far East is Everlight Chemical Industrial Corporation in Taiwan.

Intermediates account for about 30%, or $3 billion of the total market. Their prices range from about $5 to $25–$30/kg. The starting materials are mostly aromatics, particularly benzene, naphthalene, and anthracene. Thus, the well-known "letter acids" are substituted naphthalenes; for instance, H-acid is 1-amino-8-hydroxynapthalene-3.6-disulfonic acid. They are produced in simple dedicated, or at best multiproduct, plants. In its custom manufacturing offering (!), DyStar lists, among others, the following reactions: acylation, diazotization/azo coupling, halogenation, hydrolysis, nitration, reduction, and sulfonation/chlorosulfonation. Most of them are carried out in aqueous solution.

R&D focuses on product formulation and application. The development of new molecular entities has practically come to a halt.

> Examples of this trend are inks for inkjet printers. Global consumption was 4.2 million metric tons in 2008, with a market value of $19.5 billion. The pigments used are conventional: Food Black 2 for black; Direct Blue 199 for cyan; a combination of Direct Yellow 86 and 132 for yellow, and a combination of Acid Red 52 and Reactive Red 180 for magenta. The know-how is contained in the formulation. It consists of deionized water, 60–90%; water-soluble solvent (e.g., propanol, 5–30%); colorant, 1–15%; surfactant, 0.1–10%; buffer, 0.1–0.5%; biocide, 0.05–1%), and other additives (chelating agent, defoamer, solubilizer, etc.) <1%.

The situation is somewhat more positive for pigments. Effect pigments represent a category of mid to high value products. Eighty thousand tons were sold in 2009, generating sales of about $1 billion. Seventy percent of the demand comes from the coatings market (particularly the automotive sector), plastics (mainly packaging) account for 19%, and inks 5%. Infrared reflecting or absorbing pigments are used to create "energy-efficient" roof tops and green houses.

The leading producers are BASF, Merck KGaA, Clariant, Altana, Dainippon Ink & Chemicals, and Toyo Inks.

11.4.5   Electronic Chemicals

Humanity spends much the same amount on electronic goods, $800 billion, as for pharmaceuticals. The chemical industry provides most of the hardware materials. *Electronic chemicals*, an approximately $30 billion market, are attractive because the manufacturers of electronic goods are not backward integrated with regard to chemical synthesis; the merchant market for sophisticated fine chemicals is quite large and growing. Liquid crystals (LCs) are a prominent example (see Table 11.10). Because both the ongoing substitution of cathode ray tube by LC displays (LCDs), respectively thin-film transistor–liquid crystal displays (TFT-LCDs) and the shift to bigger screens, sales are expected to exceed $100 billion in 2010, four times the 2005 figure. The leading LC manufacturer, Merck KGaA, has a market share of about two-thirds. Sales in 2008 were $1.26 billion and the return on sales for the business was 44%, well ahead of that for its pharmaceutical operations. Merck's production know-how, particularly in twisted nematic cell LCs, is well covered by IPR and has enabled it to erect high entry barriers. In the backpanel segment for LCDs, the demand for optical-grade poly(methylmethacrylate) organic glass is also growing rapidly. Evonik-Degussa, the leading poly(methylmethacrylate) producer, operates a 40,000-ton/year facility with its JV partner Forhouse, Taiwan. The largest consumption of electronic chemicals is for Silicon Wafers (approximately $20 billion). With the surge in demand for the key starting material, polysilicon, for solar panels, it is expected to surpass that of the semiconductor industry.

The DuPont Zyron portfolio of high-purity etchants for silicon-based dielectric films comprises di- and trifluoromethane, octafluorocyclobutane, and nitrogen trifluoride. The latter are also used as chamber-cleaning gases.

New opportunities for use of fine chemicals in electronic chemicals emerge at a fast rate. Within LCs, inkjet systems are being introduced for dispersing pigments, such as BASF's diketopyrrolopyrrole reds (see Table 11.10). While the substitution of cathode ray tube displays by plasma and liquid crystal displays is in full swing, new types of panels are emerging. They are based on organic light-emitting diodes (OLEDs) and polymer organic light-emitting diodes (POLEDs). With a share of about 95%, OLEDS dominate the market. They are based on small-molecule materials, including substituted perylenes, quinolines, quinacridones, or polymers including polyfluorene and poly *p*-phenylene vinylene. The leading enterprises in the fledgling OLED and POLED markets are—again—Merck KGaA, BASF, Germany, and Sumitomo, Japan as traditional chemical companies, and Cambridge Display Technologies, UK and Universal Display Corp., USA as newcomers. As OLEDs have more design flexibility than traditional incandescent and fluorescent bulbs, and are expected to beat them in output (with 50 lm/W already exceeding conventional bulbs by a factor of 3 and on par with fluorescent tubes) and lifetime, they could eventually substitute for both of them. This could mean a double-digit dollar market. In order to gain wider acceptance, POLEDs must increase lifetime and efficiency. Poly-(3,4-ethylenedioxythiophene)/polystyrenesulfonate, from Bayer's Starck unit, marks a big progress as compared with the originally used polyanilines. Also, Air Products and Plextronics work on new materials that will help POLED commercialization.

An ultimately large demand for electronic fine chemicals could arise, if the performance of organic photovoltaic cells, which are cheaper to make than their inorganic counterparts, could be substantially increased. Until recently, the light-to-electricity conversion efficiency of the best performing material, a combination of a phenyl-butyric acid substituted C_{61} fullerene (phenyl-C_{61}-butyric acid methyl ester [PCBM]), a fullerene derivative, and poly(3-hexylthiophene)(P_3HT) attained merely 5%. By using more elaborate materials, the organic photovoltaic start-up company Solarmer Energy, USA expects to reach 10% by 2011.

Other new developments that will create a demand for new fine cheamicals are identification systems with radiofrequency identification tags and "smart" cards, using technologies such as organic photonic sensors. They are expected to capture a large share of the market within the next 10 years. Flexible lenses based on electroactive polymers (EAP) could eventually substitute traditional glass lens systems in photo cameras. Last but not least, a market for polymer photovoltaic solar cells is expected to gain importance.

11.4.6 Flavors and Fragrances (F&F)

F&F appeal to the palate (i.e., taste) and nose (i.e., smell), respectively. Accordingly, flavors are used in food and fragrances, in the cosmetic industries. The products originally were natural extracts from, for example, lavender, cultivated in large areas in Grasse, France; roses, grown in Bulgaria; or vanilla, grown in Madagascar. Nowadays, natural products originate mainly from China. They currently account for only 10% of the total market and are used for high-end applications, such as couturier perfumes or premium food. In order to benefit from the reputation of the term "natural," synthetic aroma chemicals are sometimes referred to as "natural identical."

Whereas natural vanilla flavor from beans (recognized by the "black dots") is used in premium ice creams, soft drinks are flavored with synthetic vanillin. Natural vanilla contains other flavoring agents as well. Thus, F&F is the only segment of the chemical industry where impurities add to the quality—and the price—of a product!

The global F&F market amounted to approximately $20 billion, at manufacturer prices, in 2009. Retail sales would give substantially higher numbers. The top 10 F&F companies add up to about two-thirds of this figure. They comprise Givaudan, Switzerland, sales (2009) $3.8 billion; Firmenich, Switzerland, $1.9 billion; International Flavors & Fragrances, USA, $2.3 billion; Symrise (merger of Dragoco and Haarmann & Reimer), Germany, $1.9 billion; Takasago, Japan, $1,4 billion; Sensient Technologies, USA, $1.2 billion; T. Hasegawa, Japan; Mane F&F, France; Danisco, DK; Frutarom, Israel.

Flavors account for the largest share, approximately 40% of the market, followed by fragrances, 35%; aroma chemicals, 13%; and essential oils, respectively natural extracts (see text above), 12%. Aroma chemicals, that is, the fine chemicals used in the F&F formulations, are produced both for captive use and for the merchant market (see Table 11.12).

About 3000 different molecules are used in the F&F industry. Approximately half of them, mostly terpenes, are included in the FEMA (Flavor and Extracts Manufacturing Association) and FDA lists. However, only a few hundred compounds are used in larger-than-ton quantities. Terpenes constitute the vast

TABLE 11.12 Major Aroma Chemicals

Category		Aroma Chemicals	
Name	Volume/Value	Name	Production Volume
Benzenoids,	88,000 MT	Vanilline	9000 MT
naphthalenoids	$675 million	2-Phenylethanol	9000 MT
		Benzaldehyde	8500 MT
		Phenyl methyl propional	8000 MT
		α-Hexyl cinnamaldehyde	7000 MT
		etc.	
Terpenes	75,000 MT	(–)-menthol	20,000 MT
	$820 million	Linalool and esters	10,000 MT
		Geraniol/nerol and esters	6500 MT
		α-Terpineol and esters	4500 MT
		Citral	4500 MT
		etc.	
Musks	14,850 MT	Polycyclic musks	10,000 MT
	$340 million	Nitromusks	N/A
		etc.	
Various	22,000 MT	Aliphatics (e.g.C_1–C_{18} acids)	N/A
	$395 million	Alicyclics (e.g., furanes)	
		Heterocyclics (e.g., pyrazines)	
Total	200,000 MT		
	~$2200 million		

Source: Adapted from market survey *Flavors and Fragrances*, SRI Consulting, Menlo Park, CA (2005).

majority of the aroma chemicals. They are obtained either by partial synthesis, starting from α- or β-pinene, or by total synthesis. There are two main industrial processes for the pivotal intermediate, 2-methyl-2-heptenone (I). The first starts from acetylene and acetone to form methyl butynol. Hydrogenation yields 3-methyl-1-buten-3-ol, which is condensed with diketene or methyl acetoacetate (Carroll rearrangement) to (I). Alternatively, (I) can also be obtained by reaction of methyl butenol with isopropenyl methyl ether, followed by Claisen rearrangement. The second one starts from isobutylene and formaldehyde. The formed 3-methyl-3-buten-1-ol reacts with acetone to yield (I). Compound I can be further converted to important aroma chemicals, such as geraniol, dehydrolinalool, and methyl ionones. Dehydrolinalool (II) is also used as intermediate for vitamins A and E. Chirality plays a big role in aroma chemicals. Takasago developed the first industrial-scale process using a chiral (BINAP) catalyst for *l*-menthol. Also BASF and Symrise, the two other producers of synthetic *l*-menthol, use chiral catalaytic hydrogenation steps. Yet, more than half of the global production is natural *l*-menthol, derived from the mint plant *Mentha arvensis*. Like Takasago for *l*-menthol (see Section 11.4.3),

Firmenich uses a BINAP catalyzed process for the manufacture of its Hedione (3-oxo-2-pentyl cyclopentane-1-acetic acid methylester) jasmine note. The F&F companies start their production typically from commercially available intermediates. Apart from the usual specifications, they also must pass customers' olfactory tests, which add a certain element of uncertainty to the business transactions. Aroma chemicals are typically both used internally and sold to third parties. The structure of dehydrolinalool is

Dehydrolinalool
(II)

As the art of creating a new fragrance—and, mutatis mutandis, a new flavor—consists in finding a new combination of existing aroma chemicals, perfumers play a major role in new product development. As the structure–odor relationship still is not well understood; this continues to be a trial-and-error approach. In R&D, the analytical departments play an important role in identifying aroma components. Thus, several hundred constituents have been found in coffee.

The future demand for aroma chemicals is expected to follow the general economic development.

11.4.7   Food and Feed Additives

Food and feed additives, also known as *dietary supplements*, are minor ingredients added to improve the product quality. Most commonly, the effects desired relate to color, flavor, nutritive value, taste, or stability in storage. The market sizes are estimated to be $20–$25 billion each for food and for feed additives, respectively. The major customers for the food additives are the big food companies Ajinomoto, Danone, Kraft Foods, and Nestlé, mentioned at the beginning of Section 11.4. With the exception of Ajinomoto, these companies are little backward integrated. As they prefer to use natural ingredients rather than synthetic ones, they are not very important prospects for the fine chemical industry. Premixers, that is, enterprises, which prepare ready-to-use mixtures of nutrients for the farmers who raise cattle, pigs, and chicken, are the main users of feed additives.

Food and feed additives do not stand back with regard to the diversity of products. They extend from minerals, mainly calcium, phosphorus, and potassium, to amino acids, terpenes (e.g., carotenoids), vitamins, and natural spices. All in all, there are several hundred individual compounds used. Saffron, made from the stigmas of the saffron crocus flower, is the most expensive product. The yearly production amounts to about 300,000 kg, and the spice is retailing for about $2500/kg. Amino acids play a big role; the largest product is monosodium glutamate (MSG), with a yearly production of 1.5–2 million tons and a price of about $2.30/kg, followed by L-lysine (850,000 tons/$1.50/kg), D,L-methionine (700,000 tons/$3/kg), L-threonine (85,000 tons/$3.40/kg), and L-tryptophane (1750 tons/$24/kg). Major producers of amino acids are Ajinomoto and Kyowa Hakko, Japan, and BASF and Evonik-Degussa, Germany. Chiral amino acids are mostly produced by fermentation. Also, many vitamins are produced more economically by fermentative processes than by traditional chemical synthesis. The vitamins are used to enhance the nutritional value of food and feed. The major producers are Adisseo, France (vitamin E only); BASF, Germany; DSM, The Netherlands; Lonza, Switzerland (vitamin B_3 only); Takeda, Japan, and a host of Chinese companies, for example, Shijiazhuang Pharmaceutical Group, Zhejiang Medicine, a.o. the world's number 1 Vitamin C producer (capacity 30,000 MT). The global production volumes correlate more or less with the Recommended Daily Allowances (RDA). For humans, the five vitamins with the highest RDA are Vitamin C (ascorbic acid), 60 mg; Vitamin B_3 (niacin), 18 mg, Vitamin E (tocopherol), 10 mg, Calcium pantothenate, 6 mg, and Vitamin B_2 (riboflavin), 1.6 mg. Correspondingly, Vitamin C enjoys the highest production volume, about 110,000 mtpa, but suffers also the lowest price, approximately $5/kg. China is slowly becoming the major world supplier. The global production capacity for Vitamin E is 75,000 mtpa, followed by niacin with a production volume of about 40,000 mtpa. The major producer with a share of 60% is Lonza. An overview of important food and feed additives is given in Table 11.13. Apart from the six categories listed in the table, there are also flavors (see Section 11.4.6, Flavors and Fragrances), medications (see Section 11.3, Animal Health Industry), buffers, colorants, growth-promoting antibiotics, hormones, microbial cultures, minerals, pellet binders, and preservatives.

Nutraceuticals, derived from nutrition and pharmaceuticals, refer to food additives claimed to provide a pharmacological benefit. As the movement from treatment to prevention stimulates the market, they are enjoying an increasing popularity. The global market is expected to grow at a rate of 7.5% per annum, from $135 billion in 2009 to $200 billion in 2015, of which the active ingredients, however, account only for a small fraction. Examples of nutraceuticals, also referred to as dietary supplements, are anticyanins, antioxidants, such as β-carotene, hypericum, as antidepressant, isoflavonoids to improve arterial health, or sulforaphane as a cancer preventative. Also, L-carnitine, flavonoids, resveratrol, and ubiquinone are widely used.

A complete listing is published by the U.S. Dietary Supplement Health and Education Act (DSHEA).

Most of the food and feed additives are commoditized. This is also the case for the artificial sweeteners, aka sugar substitutes or low-calorie sweeteners. The global market of about $3.2 billion comprises both synthetic and

TABLE 11.13   Major Food and Feed Additives

Category	Food and Feed Additives
Name	Name
Amino acids	MSG, monosodium glutamate
	D,L methionine, L-lysine
	L-threonine, L-tryptophan
Antioxidants	Ascorbic acid
	BHA, butylated hydroxyanisole
	BHT, butylated hydroxytoluene
	Tocopherols
Artificial sweeteners	Acesulfame-K, aspartame, cyclamate, saccharine, sucralose
Carotenoids (tetraterpenes)	β-carotene
	Axanthin
Preservatives	C_1–C_3, citric, lactic, sorbic, and benzoic acids and salts
Vitamins	Fat-soluble: A, D. E, K
	Water-soluble: C, B_1, B_2, B_3,
	B_6, B_{12}, H, folic acid, Ca pantothenate

natural products. Splenda is the top-selling product. Its active ingredient, sucralose (600×) had been discovered by Tate & Lyle in 1976. It enjoyed a spectacular revival after Johnson & Johnson's formidable marketing machine begun selling it. Following suit in sales of artificial sweeteners are saccharin ("Sweet'N Low," 550×), aspartame ("Canderel," "Equal," and "NutraSweet," 200×), and acesulfam K ("Sunnett," 200×"). There is also a long list of natural sweeteners. Examples are Brazzein (800×), Mannitol (0.5×), Stevia (250×), and Xylitol (1×). The numbers in brackets are the "sweetness intensity," whereby sucrose ≡ 1.

11.4.8   Specialty Polymers

Specialty polymers, also known as *engineering plastics*, which exhibit particular chemical and mechanical properties, are more expensive and are produced in smaller quantities than the large-volume/low-price commodity polymers, such as polyethylene, polypropylene, polystyrene, and poly(vinylchloride), which are household names. Specialty polymers are not well-known to the general public. They are used primarily in the electronic and aerospace industries, where very demanding electrical, respectively mechanical, properties are required. A large portion of the added value in specialty polymers lies in the polymerization process itself. An extreme case is the expensive "organic metal" polyaniline, which is made from the commodity aniline. Fluorinated poly(ethylene/propylene) has the largest market (see Table 11.14). As the producers, Daikin, Japan and DuPont, USA, are fully backward integrated, this polymer does not offer business opportunities for the fine chemical industry.

TABLE 11.14   Specialty Polymers

Type	Market Size (2005)			Application/Backward Integration of Producers
	Volume (tons)	Unit Price ($/kg)	Turnover ($ millions)	
Thermoplastics				
Polyether ether ketones (PEEK)	5000	85–100	≈ 450	Aerospace, electronics, medical implants
Fluorinated ethylene/ propylene	30,000	60	1800	Wire and cable insulation backward integrated
High-temperature thermoset resins				
Non-melting polyimides	8000	30	240	Aerospace and rigid electronic parts partly backward integrated
Cyanates	2000	30	60	Aerospace and rigid electronic Some producers buy advanced intermediates
Electronic coatings (thermoset and thermoplast)				
Benzocyclobutene polymers	5[a]	100[a]	0.5	Wafer coatings Producer is backward integrated
Polyimides	8000	5–250	≈ 100	Aerospace and rigid electronic parts Some producers buy advanced intermediates
Liquid crystalline polymers (smectic and nematic)				
Aramides, aromatic polyesters, polypenzazoles	25,000	20	500	Electronic connectors and precision thermoplast parts; some producers buy advanced intermediates

[a] 30% solution.

Note: A number of branded specialty polymers and their manufacturers are listed in Table 11.10.

Polyimides are the most versatile specialty polymers in terms of applications. Starting materials are dianhydrides, such as pyromellitic dianhydride or bis(phenyltetracarboxylic acid dianhydride) on the one hand, and aromatic diamines, such as phenylenediamine or diaminodiphenyl ether on the other hand. Polyetheretherketones (PEEK) are engineering plastics with extraordinary mechanical properties, outperforming steel on a strengths-to-weight ratio. They are obtained by condensation of aromatic dihalides, for example, di-(p-chlorophenyl) ketone with bisphenolate salts. The semi-crystalline material is used in bearings, piston parts, and pumps, and is also considered for implants. Producers are Changchun Jida High Performance Materials, China; Evonik-Degussa, Germany; Solvay Advanced Polymers, USA and India; and Vitrex, UK.

CHAPTER 12

Marketing

12.1 ORGANIZATION AND TASKS

Large enterprises differentiate between marketing, which is viewed as "all activities of a company dealing with the market" on the one hand, and sales, which comprises the commercial interaction with the customers, on the other hand. In the fine chemical industry, which is mainly midsized and rather lean with regard to staff positions, marketing is understood primarily as an operational function within the business units. Thus, only the high-risk, future-oriented task of developing business for new products in new, hitherto unknown markets (see rectangle no. 3, i.e., the "forbidden zone" in Fig. 12.1) is assigned to Corporate Development, that is, a function outside the profit and loss (P&L) responsibility of a business unit (b.u.). As the fundamental rationale for the organizational structure of the b.u.'s is customer orientation (see Fig. 8.3 in Section 8.3), marketing within the b.u.'s as well is aligned to the markets served, that is, pharma, agro, and nonlife science industries, as the case may be. The marketing functions cover sales, business development, pricing, and promotion.

> Because its purpose is to create a customer, the business enterprise has two, and only these two, basic functions: marketing and innovation. Marketing and innovation produce the results; the rest are "costs."
>
> —Peter Drucker, *Marketing*, 1973

In the business unit pharma, which in most cases is the largest one, a clear separation between sales persons in charge of active pharmaceutical ingredients (APIs)-for-generics on the one hand, and those responsible for custom manufacturing (CM), resp. exclusives, on the other hand, is recommended. In broad terms, the former are responsible for the combinations "existing products/existing markets" and "existing products/new markets," the latter for "new products/existing markets." The term "markets" is understood as "customers" in this context. Upon further scrutiny, it becomes evident that

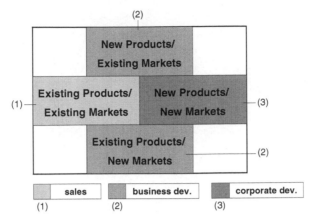

Figure 12.1 Product/market matrix.

TABLE 12.1 Characteristics of Fine Chemicals for Originator and Off-Patent Pharmaceuticals

	Exclusives (Custom Manufacture)	Generics (API-for-Generics)
Business model	Project driven	Product driven
Products	Advanced intermediates	End products (APIs)
Pricing	"Buttom-up"	Market price
Distribution channels	Direct	Agent
Customers	Ethical pharma companies	Generic pharma companies
Competitors	Captive production 1–2 suppliers/product	Asian producers
Competitive advantage	Project management	Price, quality, DMF[a]
Origin of know-how	Customer	Supplier
Technical assistance	Close cooperation	Sporadic
Legal assistance	Sporadic	Intensive
Production planning	On order	Min./max. stock

[a] DMF, drug master file.

the two tasks differ substantially. As shown in Table 12.1, business aspects prevail in API-for-generics, project management/technical aspects in CM.

In large fine chemical companies, the creation of a third marketing function, business development, is worthwhile to be considered. It is positioned besides API-for-generics and CM, which essentially are sales functions. Whereas they have the task of optimizing profits at present (i.e., meeting the budget targets for the current year); business development is accountable for future sales and profits, in particular, the establishment and realization of elements becoming increasingly important during the realization of a business plan. Business development generates ideas for new products and services and follows them

TABLE 12.2 Sales and Business Development—Specifics

Differentiator	Sales	Business Development
Time horizon	Present (budget)	Future (5-year business plan)
Product category	Catalog or standard[a]	Exclusive (CRAM)
Customers per product	Many (nonexclusive)	One (exclusive)
Unique Selling Proposition (USP)	Product price, quality, reliability	Service project management
Qualification of salesperson	Business school	Chemical degree (+ business acumen)
Pricing	Market price	Individual ("bottom-up")
Process ownership	Seller	Customer (technology transfer)
Prime internal contact	Production	R&D
Production site	Plant	Lab → pilot plant → plant
Specifications	From seller	From customer
Technical assistance	Sporadic	Close cooperation
Production planning	Max./min. stock	Order
Distribution channel[b]	Local office or agent	Direct

[a] Listed in the company's websites and product brochures.
[b] See Section 12.4.

through to successful realization. As such, it mirrors the supplier selection of the customer. This typically is a four-step process, namely: (1) internal assessment (desk work) →(2) request for information (RFI), typically done by submitting a questionnaire →(3) audit of shortlisted candidates →(4) request for product (RFP).

The important differences in the functional specifications of sales and business development are described in Table 12.2. For a job description for a business development manager, see Appendix A.5. Both sales and business development report to the head of marketing in the b.u., respectively of the division head in larger companies. In many companies, there is an intrinsic lack of understanding between business development and R&D. Business development pretends that "R&D develops primarily new products for which there is no market," whereas R&D's position is: "Our resources do not allow the development of a suitable synthesis for the kind of new products that business development proposes." In order to overcome this impasse, the creation of a "new product committee" has proved very useful. The committee has the task of evaluating all new product ideas following a standard checklist (see Appendix A.2). It decides whether a new product idea should be taken up in research, and thus becomes a *project* (see also Section 6.2, Project Initiation), and whether an ongoing research project should be abandoned, a frequently neglected duty. A warning signal would be if the chance of both a commercial and a technical process diminish continuously over a period of several months! The committee is chaired by the head of marketing; full-time members are experienced researchers and business development managers. Ad hoc members are specialists from controlling, production, QC/QA, and the legal department. The committee should meet every 2–4 weeks.

DEFINITION OF BUSINESS DEVELOPMENT

Develop profitable new business, by

- *Identifying* potential customers on the basis of a fit between their needs and your company's competences
- *Evaluating* the attractiveness of new business opportunities
- *Offering* targeted new products and services
- *Initiating, developing*, and *completing* joint projects, building on your company's capabilities and resources

---

Whether a business development manager should transfer new products, when they become mature (e.g., once a supply contract is about to be concluded) to his colleagues in API-for-generics or—the more frequent case—in CM is a controversial issue. From the perspective of customer orientation, the product stewardship should not be transferred, particularly if the responsibility for the procurement remains with the same person at the customer's company. In terms of having "the right person at the right place," a business development manager should concentrate on its primary mission, namely, the acquisition of new business. The same problem arises within R&D: Should a researcher be confined to the laboratory, or move with a new product project from laboratory to pilot plant, and ultimately to the plant? As a compromise, the lab scientist who developed the new product process in the lab introduces it into the pilot plant and is responsible for demonstrating the feasibility on this scale. A limited competition and substantial unmet needs would describe a favorable working environment for business development. For CM, neither is the case nowadays. Despite a trend to more outsourcing by the life science companies, the number of fine chemical companies and the global Good Manufacturing Practice (GMP) fine chemical production capacity have increased more rapidly than demand over the past 10 years (see Chapter 10).

The development of new products for new markets (e.g., electronic chemicals for a fine chemical company so far dealing only with life sciences) is the task of corporate development. It deals with activities outside the authority of the operating b.u.'s or divisions. Apart from "new products for new markets," it also looks after corporate identity, joint ventures, mergers and acquisitions, and divestitures. A further option for structuring of the marketing function is the distinction between small and big molecules. This is only recommended in business development of large fine chemical companies. Here, the assignment of a biochemist having a specific knowledge of the intricacies of biotechnology can be advantageous. In the other functions, the rules of the game are the same, regardless of the size of the molecules.

An interesting variety of the organization scheme is the combination of the positions of head of marketing and R&D in one and the same person. It eliminates the abovementioned friction. This structure is particularly suitable for small to midsize fine chemical companies, where the total number of researches and business development managers does not exceed a dozen. The ideal candidate for this position is a scientist with several years of experience in industrial R&D and pilot plant production, plus a commercial flair.

In order to determine the adequate number of sales and business development managers needed, it is helpful to refer to the—present, respectively future—sales, which each individual can manage. As a rule of thumb, this number is about $10–$15 million per year. Thus, a fine chemical company that presently has $100 million in sales would need 7–10 marketing managers (sales and business development combined).

12.2 TARGET PRODUCTS AND SERVICES

The products and services offered by the fine chemical industry fall into two broad categories: (1) "exclusives" and (2) "standard" or "catalogue" products. Service-intensive "exclusives," provided mostly under contract research or CM arrangements, prevail in business with life science companies; "standards" prevail in other target markets.

As mentioned before (see Section 11.1), the life cycle of a specialty chemical lends itself for the identification of business opportunities. This is particularly the case for pharmaceutical fine chemicals (PFCs) but, *mutatis mutandis*, it also applies for agrochemicals, veterinary drugs, and other target products. During the life cycle of a drug, there are four "windows of opportunity" for identifying target products (see Fig. 12.2). The number of possible product candidates, the chances of success of the drug to make it to the marketplace, and the chances of a specific fine chemical company to get the business change substantially along the product life cycle. The four opportunities, namely pre-clinical, development phase II/III, maturity, patent expiration, span a period of approximately 20 years, during which the drug is under patent protection. At the beginning, there is a huge number of lead compounds, typically 5000–10,000, in the R&D laboratories of the world's pharmaceutical industry ... of which only 1 (!) makes it to the marketplace and therefore ultimately requires substantial volumes of APIs. At entry gate number 1, the emphasis is on quickly producing small-scale samples by simple processes without economical or ecological considerations. This "quick and dirty" approach is the domain of the contract research organizations (CROs; see Section 2.2). During the subsequent clinical development, chances of a commercial success become gradually more tangible (see Table 12.3). This is where CM kicks in. Phase II

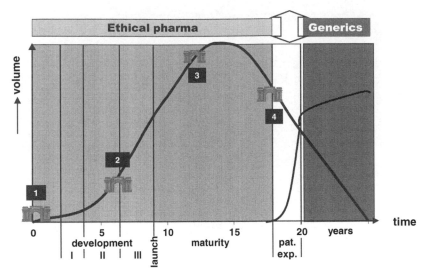

Figure 12.2   Product opportunities along the drug life cycle.

TABLE 12.3   Drug Development Phases

Development Stage	Preclinical[a]	Phase I[b]	Phase II[c]	Phase III[d]	Phase IV[e] Prelaunch
Number of drugs in development	≈5000	1000–1500	1500–2000	≈500	—
Rate of success	<<1%	10–15%	20–25%	50–70%	90%
Volume needs	10^1 kg	10^2 kg	10^3 kg	10^3–10^4 kg	10^4–10^5 kg
Cost[f] ($ million)	80–120	60–120	50–150	150–300	20
Timing (years)	3–6		← 6–8 →		0.5–2
Key activities	Contract research	Synthesis sequence selection, proof of concept	Development of second generation process		Regular production

[a] Lead discovery; *in vitro* testing.
[b] Small-scale clinical trials with healthy volunteers to determine safety, tolerability, and ADME (absorption, distribution, metabolism, excretion).
[c] Trials with 100–500 volunteer patients to determine the drug's efficacy in treating the disease it is intended to cure.
[d] Double-blind trials with hundreds to thousands of patients suffering from the relevant disease to get more information on efficacy, safety, and side effects; preparation of registration documents.
[e] Prelaunch marketing and production of stock; registration.
[f] Total drug development cost per successful launch $1.4 billion.
Source: Reference 5.M. Bloch et al., Pharma leaps offshore, *McKinsey Quarterly Newsletter*, July 2006.

of clinical development constitutes a pivotal point. Pharmaceutical companies lock in the final route to be used for manufacturing their new APIs.

For a systematic search for product candidates, the single most important information sources are IMS Health for pharmaceuticals, Agranova for agro, and Vetnosis for veterinary drugs (see Appendix A.1). Useful information for both existing and developmental products can be found on the websites of the pharma and agro companies. Particularly the larger companies publish ample details, including a detailed description of their new product pipelines and the sales figures for their major commercialized products. A deeper insight into new product projects within the life science industry is provided by a number of comprehensive project databanks. A selection is also listed in Appendix A.1.

12.2.1 Exclusives: Custom Manufacturing

CM constitutes the "Königsdisziplin," that is, the most prominent activity of the fine chemical industry. CM is the antonym of outsourcing. Because of their long-standing customer relationships, their reliability, and the experience in safeguarding the intellectual property (IP) interests of its customers, it is has been the stronghold of the Western companies since its inception in the mid-1970s.

DEFINITION OF "OUTSOURCING"

The transfer an industrial activity—production or service—that so far was carried out in-house to a third party through a contractual arrangement.

Note: If outsourcing is done with an overseas partner, it is also referred to as *offshoring*.

In CM, a specialty chemicals company outsources the process development, pilot plant, and, finally, industrial-scale production of an active ingredient, or a predecessor thereof, to one, or a few, fine chemical companies. The IP of the product, and generally also the manufacturing process, belong to the customer. The customer–supplier relationship is governed by an exclusive supply agreement (see Section 12.7). At the beginning of cooperation, the customer provides a "tech package," which in its simplest version, includes a laboratory synthesis description and safety, health, and environment (SHE) recommendations. In this case, the whole scale-up, which comprises a factor of about 1 million (10 g $\rightarrow$ 10 ton quantities), is done by the fine chemical company.

Toll manufacturing is a variety of CM. As in CM, the know-how is provided by the customer, frequently a backward-integrated agrochemical company. The

TABLE 12.4 Generics Patent Legislation (Examples)

Name of Law	Content
SPC = Supplementary Protection of Certification	Allows for a patent extension of approximately 5 years or longer
Roche/Bolar Provision	Consents the preparation of a.o. "experimental quantities" of API-for-generics prior to patent expiration
30 months postponement of FDA approval	Takes effect if an originator pharma company sues a generics company for patent infringement
180 days market exclusivity	Granted to the first generics company challenging the validity of a brand name
Loophole in the Hatch-Waxman Act	"Authorized Generics" is a tactic increasingly used by branded pharma companies, whereby they relabel their *branded product as a generic and introduce it during the 180-day exclusivity*

starting material is generally supplied free of charge. The toller, therefore, must take great care to reach the agreed-on yield. Also, total sales revenues of a toll manufacturer are much lower (typically less than half) than those of a custom manufacturer, which purchases the starting materials. Contrary to CM, a full-fledged industrial-scale process is provided, meaning that technology transfer occurs on a one-to-one scale. Only a minor involvement of R&D is necessary. Adaptation of the process to the specifics of the supplier's plant is a task of chemical engineers rather than R&D chemists. This is, for instance, the case in a solid–liquid separation step, if a switch has to be made from a centrifuge to a filter dryer.

Custom and toll manufacturing account for the majority of business between fine chemical and life science industries. The specialty chemicals industry outside life sciences sources mainly standard products.

Within CM for pharmaceutical companies, entry gate number 2, when the new drug is in Phases II and III of development, is most attractive for established fine chemical companies (see Fig. 12.2). Here, the so-called first-generation manufacturing process for the new API is determined ("frozen-in"), and the make-or-buy decision (i.e., in-house production or outsourcing) is taken by the potential customer. At the same time, the chances for a successful launch are also becoming tangible. If the outsourcing decision is positive, pharmaceutical companies generally entrust CM projects at the Phase II/III development stage only to approved suppliers. Newcomers must start their business development effort already at Phase I. In order to do so, fledgling fine chemical companies offer a "one-stop shop," or contract research and manufacturing (CRAM) service (see Chapter 2). This allows them to build up a reputation by supplying samples early on. The price at which a daily dosage can be sold is a pivotal element in the new drug project. Particularly for high-dosage drugs, this puts pressure on the price of the embedded API. Entry gate number 3 is located close to the peak level of the drug production volume. Drugs that are introduced in the same therapeutic class by competitors start to negatively impact on the market share at this point. COGS (costs of goods sold) become a major concern, and therefore, more cost-effective second-generation processes are urgently needed. This presents a chance for a new supplier. Entry gate number 4 is linked to the decline phase of the life cycle, when patent expiration is approaching. In order to stave off generics competition, it becomes a question of survival for the originator company to develop *the* most economic process. Business opportunities surface with both (1) the ethical pharma company that holds the patents, and attempts to keep at least a portion of the market, for example, by creating an "Authorized Generic" (see Table 12.4) and (2) generics houses that are preparing for the launch of generic versions The forthcoming generic competitors are vying for market share—and are also looking for competitive suppliers. Volume-wise, this gateway is attractive, but the competition for the business is intense.

The chance of a particular fine chemical company to build up a rewarding exclusives business depends on a "two-level demand hierarchy," namely, (1) the overall outsourcing policy of the customer base, primarily the pharmaceutical industry (this is discussed in Section 10.2, Demand) and (2) the fine chemical company must succeed in acquiring a particular business. This depends on its abilities to meet the customer's needs. As the "hardware" of the top-tier fine chemical companies does not differ very much nowadays and complies with high standards, the success factors have shifted to the "software," such as communication, role definition, and empowerment of team members. It is imperative to establish a good working relationship and set up an infrastructure for clear communications. Too often, the rush to begin work skips this step; the result is delays and misunderstandings that can dog the working relationship for years and reduce the opportunities for achieving high quality. Relationship building is especially likely to falter when the distance between the partners is great, or where differences of culture (corporate or country) and language are large.

Best practices include

• Frequent, honest, thorough communication	• Defined role for each firm, department and team member
• Structured written system to communicate	• Clarity of authority
• Written quality agreement, meeting the partner's needs	• Periodic debriefing review of exceptions, communication, and expectations
• Nomination of a steering committee	• Flexibility in face of changed situations
• Training and briefing of all team members	• Shared ownership of successes

The selection criteria for outsourcing partners, as established by a major pharmaceutical company, are listed in Appendix A.6. Those fine chemical companies that best comply with the 28 criteria obviously have the best chance to become suppliers. The customer's choice is typically based on an extensive audit of a short-list of prospective suppliers.

12.2.2 Nonexclusives: API-for-Generics

Many fine chemical companies are active in both CM and API-for-generics, albeit with quite different proportions. The divergences are driven mainly by the history of the patent legislation in a given country. Companies located in countries having adhered to international patent legislation for a long time, particularly those in Germany, France, The Netherlands, Switzerland, the United Kingdom, and in the United States, are mainly active in CM. In southern Europe, Malta benefits from the fact that many pharmaceutical companies did not bother to apply for product patents. Also companies operating in Italy and Spain, having benefited from a privileged patent regime for many years, are involved mainly in API-for-generics [4]. Product patents were enacted in Italy only in 1978 and in Spain in 1992. The same applies for China and India, which joined international patent legislation in 2005. Before, they were at liberty to produce and sell APIs for patented drugs in their home markets, as well as in other non-protected markets. This is the main reason why the fledgling fine chemical companies in these countries started their activities with API-for-generics. Globally, the four largest API-for-generics producers are China, with a market share of 30%, followed by Italy, 20%, India, 13%, and Spain, 7.4%.

Custom manufacturing PFCs for an ethical as opposed to making API-for-generics for a generic pharmaceutical company are *two different businesses*. The only common denominators are the hardware, multipurpose plants, which are shared, and the chemical process technology. The divergences are listed in Table 12.1. CM is a *project* business involving a close cooperation between customer and supplier, and extending from the supply of samples for the first clinical trials all the way to pilot plant and large-scale industrial manufacture of PFCs for commercial drugs. API-for-generics is a *product* business. Apart from some legal and regulatory intricacies, it has all the characteristics of an ordinary commercial deal. Within CM, fine chemicals of the advanced intermediates category are most important, followed by nonregulated intermediates. There is only a limited market demand for APIs, as these are usually produced in-house by the customers for quality, intellectual property rights (IPR), and tax considerations. The situation is different for PFCs going into generics, where only APIs are traded. From a production perspective, it makes little difference whether a fine chemical company manufactures PFCs for patented or off-patent drugs. Actually, the combination of CM and API-for-generics activities can lead to better capacity utilization. Whereas the production schedules for exclusive products are rigid, there is more freedom in planning campaigns for API-for-generics.

Plagued by a sluggish demand for their CM business (see Section 10.1), many fine chemical companies are considering starting or expanding their own API-for-generics business. The main justification is the brighter outlook for demand growth (see Section 16.2). There is much debate within the industry as to whether it makes good business sense to supply PFCs to both ethical and generic pharma companies, which are competitors. The answer is that it can be done, provided the following basic rule is observed: A custom manufacturer must never supply a specific API, or precursor, which it is producing for an ethical pharma company, to a generic house. This applies even if the supply contract has expired and the legal situation would allow it. The position of the pharma customer is that the custom manufacturer would unfairly benefit from the know-how gained during the cooperation period. On the other hand, there are a large number of old generics that are sourced by both ethical and generic companies, and where there is no objection to production and marketing.

The markets for API-for-generics are expanding rapidly (see Chapter 17). Also, they do not require any technology for their manufacture different from that of APIs for patented drugs. Apart from these attractive aspects, there are, however, also major challenges to be dealt with, when considering the API-for-generics option. Low prices are the raison d'être for generics. Because of the relatively low entry barriers, the competition in generics in general and their active ingredients in particular is fiercer than in CM. Asian companies are in an excellent position to compete on the basis of price. Confidentiality considerations are much less relevant. Consequently, the Chinese, Indian, and South Korean fine chemical companies have been aggressively penetrating this market. Asian companies have been investing heavily in fine chemical manufacturing capacity, equal to an estimated 33% of the worldwide total. Most of it is used for API-for-generics. According to a market study on APIs from the Chemical Pharmaceutical Generic Association in Italy [4], European firms produced about 33% of the worldwide merchant market for APIs in 2008, down from 45% in 2005, and from nearly 65% in 1990. As the customer base is different, existing customer relations are of no value. There is no customer loyalty. In most cases, agents are used as intermediaries. Also, the argument that the intrinsic added value is highest for API-for-generics has to be reviewed. It is correct that the APIs represent the last stage in the synthetic pathway, whereas in "exclusives," the last step is seldom outsourced and performed by the customer. One has to consider, however, that the APIs of modern proprietary drugs have a more complicated chemical structure. Therefore, making an advanced intermediate for a proprietary drug may require more steps, and consequently more added value, than making an API-for-generic. Last but not least, a substantial financial risk is associated with the development of a new API: The R&D expenditures alone amount to several million dollars and have to be spent many years prior to the product launch. Other expenses incurring are legal fees, filings with the authorities, and plant adaptations.

In order to be (or become) a successful player in the API-for-generics market, a fine chemical company must excel in the following six areas:

1. *Picking the Winners.* More than 2000 APIs are on the market today. In order to find a suitable candidate for its own product portfolio, a fine chemical company has to carefully screen and evaluate all of them on the basis of commercial, legal, and technical criteria. Whereas the technical evaluation can be carried out by in-house experts, third-party advice is recommended for legal and commercial aspects. In the United States, specialized agents with a profound knowledge of the generics market are Aceto, Betachem, Davos, Gyma, Interchem, SST, and Vinchem. In Europe, Arnaud Group in France, Alfred E. Tiefenbacher, Helm, and Midas in Germany, Veltronelli in Italy. Fischer Chemicals in Switzerland and S&D Chemicals and Whyte Group in the UK are well-known agents. In Japan, the big trading houses Mitsubishi Corporation, Mitsui & Co., Nagase, and Sumitomo Corporation are the most proficient and knowledgeable players. Within the technical criteria, products needing only "pots and pans" for their manufacture should be ignored in favor of those requiring one or more niche technologies, such as energetic chemistry, low-temperature/high-pressure reactions, chiral synthesis, or separation. The commercial parameters should include the sales revenues (products with either very high or very low sales should be excluded; see end of Section 12.2.2), the competitive situation, and the growth prospects.

2. *First on the Market, Respectively "First-to-File."* As market prices for generics and API-for-generics erode very rapidly after patent expiration of the originator product (see Fig. 12.3), good profits can be made only during the first months following the launch. In order to take advantage of this situation, a generic company must be ready to commercialize its product virtually the day after the patent expiration.

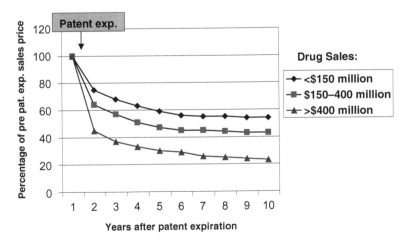

Figure 12.3 Price decrease of APIs after patent expiration.
Source: European Association of Generic Producers, IMS Health Service, CGEY, 2002.

For instance, in its 2002 annual report, Barr Laboratories mentions as major highlight of the year: "Launched and shipped 100 million generic Prozac® capsules in first 48 hours of FDA approval." KUDCO, the U.S. subsidiary of the German Schwarz Pharma, launched a generic version of Prilosec (omeprazole) on the day after the patent expiry (Dec. 20, 2002), and matured sales of $150 million until the end of the year, bringing the turnover up by 20% (!) for the full year, and YULE CATTO, producer of the API for omeprazole, reported a 30% increase of its EBITDA for 2002.

Thus, the API-for-generics producer has to prepare both the manufacturing process and the dossier for the ANDA (abbreviated new drug application) submission well in advance of patent expiry. European fine chemical companies, where the Roche/Bolar ruling does not apply (see Table 12.4), are at a big disadvantage, because development work cannot start prior to patent expiration. Therefore, they are looking for possibilities of doing at least the last step of the synthesis outside EU countries. Macao, Malta, Turkey, and even the United States are chosen as sites for the final stage of production. In contrast, when Asian companies launch their API-for-generics on Western markets, they benefit from many years of production experience gained while they marketed the products in their home markets.

3. *Intimate Knowledge of the Patent Situation.* The generics business is regulated by a large number of patent laws, some in favor of the patent holders, and some in favor of the generics companies. An insufficient knowledge of the patent situation for any given drug can trigger lengthy and expensive lawsuits. The generics company can also be obliged to withdraw the product. Four examples of the legislation are given in Table 12.4. Ethical pharma companies undertake every effort to extend the patent protection for their most important moneymakers. Over the years, they have developed a large arsenal of weapons for this purpose, on both legal and technical sides; on the legal side, patent extensions are sought by filing new patents for the manufacturing process, or for new applications (e.g., BMS's pediatric exclusivity extension for glucophage). While suing the generics companies for patent infringement, they are granted a patent extension. On the technical side, specifications for pivotal intermediates are tightened, combination drugs (e.g., Schering-Plough/ Merck's Vytorin, a Zetia/Zocor combination cholesterol-lowering drug) are developed, numerous polymorphous crystal forms—for which no claims were made in the original application—are patented (e.g., Pfizer's Sertraline), or racemic switches are used. The latter refers to the development of the single isomer form of a racemic compound. An interesting example in case is AstraZeneca's anti-stomach ulcer drug Nexium (esomeprazole), which is one of the enantiomers of the precursor Prilosec (omeprazole). After having patented Nexium, AstraZeneca sued the generics companies selling Prilosec on the basis that it was containing the—patented—Nexium active ingredient.

4. *Low-Cost Producer.* The price competition in generics is fiercer than in CM, primarily because low prices are the raison d'être for generics, and because the Far East companies have been penetrating the market aggressively. The average cost of a prescription for a generic is 30% or less that for an originator drug. Whereas confidentiality and IPR considerations constitute an entry barrier to CM, fine chemical companies in this region can fully exploit their "high-skill/low-cost" advantage in API-for-generics. Moreover, they benefit from the learning curve experience, as they were allowed to produce APIs for patented drugs under the former legislation. The new legislation applies only to pharmaceuticals whose patents expired after 2005.

An example is the AstraZeneca's cholesterol-lowering drug Crestor (rosuvastatin). Although the patent expires only in 2012, there are already about 25 manufacturers in India, including famous names such as Ranbaxy (Rosuvas) and Dr. Reddy's Laboratories (Rosvat).

Western companies should shy away from products where they can expect minor reductions of the manufacturing costs only and go for those where they can achieve breakthrough process improvements. As a result of the highly competitive business climate, prices for API-for-generics drop significantly as soon as the relevant originator product patent expires. The price collapse is a consequence of the loss of the monopoly position due to patent expiry and a lower cost base of the generics industry; the slump in market share is caused by the lack of promotional support and the launch of new drugs in the same therapeutic category. As illustrated in Figure 12.3, the price decrease is particularly dramatic in year 1 after patent expiration, where they already dropped to between 75% and 45% of the original level, and for those drugs that had achieved high sales volumes when they held the proprietary status. These drugs are obviously targeted by a large number of API-for-generics manufacturers. As illustrated in Figure 12.3, prices for APIs for drugs with sales exceeding $400 million per year in the last year prior to patent expiration drop to about 30% of the original value after 5 years and to 20% (!) after 10 years. APIs for drugs in the sales category of $150–$400 million drop to 47%, and those with sales below $150 million, plunge to 59% in the same period.

5. *GMP Culture.* For the manufacture of API, the rules for GMP must be observed. Plants for the production of APIs are subject to FDA inspection, regardless of whether they are used for patented or non-patented drugs. Compliance with GMP regulations, therefore, does not constitute an entrance barrier for fine chemical companies that thus far produced APIs for patented drugs only. However, the capability to prepare the regulatory documentation necessary for a successful ANDA (Abbreviated New Drug Application) is a key differentiator.

TABLE 12.5 API-for-Generics: What Is in for a Fine Chemical Company

Originator Drug	No. 1 Lipitor (Pfizer)	No. 101 Femara (Novartis)
Therapeutic class	Cholesterol reducing	Antineoplastic
Finished U.S. drug sales 2008	$5900 million	$360 million
API value (5%/5%)	$300 million	$18 million
First year price drop (55%/35%)	$135 million	$12 million
Drug market share drop (50%/50%)	$70 million	$6 million
Market share of fine chemical company (10%/100%)	$7 million	$6 million
Earning on sales (10%/25%)	*$0.7 million*	*$1.5* million

6. *Financial Muscle.* The developmental work for a new API-for-generic has to start 5–8 years before patent expiry of the originator drug. This means a substantial investment in both human resources and R&D funding many years before any revenues are generated.

In conclusion, going after APIs for proprietary blockbuster drugs with forthcoming patent expirations is not a valid business proposition for a fine chemical company. It would require a large upfront investment in R&D, coupled with a high risk of failure. This applies first of all to Western companies but increasingly also to their Asian competitors. This category is the domain of the leading generic companies. An example is outlined in Table 12.5. For a fine chemical company, the generic ranking number 101 in U.S. sales, Novartis' Femara (letrozole), has a profit expectation twice to that of the world's number 1, Pfizer's Lipitor (atorvastatin)—at a lower risk! Instead of vying for a small piece of the market for atorvastatin, which will become accessible in 2013, the efforts should rather be directed to find a more economic synthesis for the API for, for example, Atenolol, which is already generic. For obvious reasons, going after APIs-for-generics with small sales, typically below $100 million per year, is not attractive, either. Business opportunities exist in the areas of niche products with average sales, especially if special technologies are required, and in "continuing to make" a PFC for a proprietary drug that is becoming generic. One last open question remains: Should the marketing of a fine chemical company direct their business development on large or small generic houses? The powerhouses, such as Teva and Sandoz, obviously require large volumes, but they are backward integrated and have a tremendous purchasing power. In contrast, the small ones totally depend on sourcing from third parties, but usually require volumes below the minimum economical size of a supplier.

As biopharmaceuticals entered pharma markets much later than small-molecule drugs—as first product recombinant insulin was launched in 1982—patent expirations have occurred only recently. The generic versions of biopharmaceuticals, called *biosimilars*, or *follow-on biologics*, are now gaining importance as new type of generics. They will be discussed in Section 17.2.

12.2.3 Standard Products

Standard products, also known as "catalog products" or "building blocks," turn up first, if one browses the product pages of websites or brochures of fine chemical companies. Except for laboratory chemical suppliers (see Section 2.3), they rarely play a major role in the product/service portfolios. Production requirements are less stringent than for GMP fine chemicals. Contrary to exclusives, standard products derive from a "supply push" rather than a "demand pull" marketing approach:

- They are developed by big chemical or petrochemical companies in order to extend their value-added chain, for instance, in the sequence methanol → acetic acid → diketene → acetoacetates → γ-chloroacetoacetates → 2-aminothiazolyl acetates, or benzene → cumene → phenol → salicylic acid.
- Process technologies used primarily for the production of large-volume commodities, such as ammonoxidation for acrylonitrile, halogenation for alkyl chlorides, hydroformylation for fatty alcohols, nitration for nitro-benzene, -toluenes, and -xylenes, or phosgenation for methylene diisocyanate are further exploited for the production of more advanced intermediates.
- By-products of large industrial-scale processes are valorized; for instance, in the DuPont process for adiponitrile from butadiene, the by-product α-methylglutaronitrile is upgraded to β-picoline and further to niacinamide.
- Intermediates from synthetic pathways carried out to produce exclusive products are offered to third parties.

Other examples are acetoacetates; alkylamines and alkyl halides/acid halides; ethers; esters; chloroformates; ketones; lactames; lactones; malonates; mercaptanes and orthoesters in aliphatics; catechol/hydroquinone/resorcinol, cresidines; haloaromatics in aromatics; and coumarines, cyanuric chloride, picolines, quinolines, and thiazoles in heterocyclics.

In most cases, the products obtained by this way are commodities, as they are produced in large volumes in dedicated plants and cost less than $5–$10/kg.

12.3 TARGET MARKETS: GEOGRAPHIC REGIONS AND CUSTOMER CATEGORIES

The attractiveness of specific product categories, as discussed in the previous section, by and large also defines the attractiveness of target markets. Besides its absolute size, the pharmaceuticals market comes first because of its inherent elevated added value, the relatively high innovation rate, which leads to a steady demand for new products, and the manageable number of potential customers. The attributes for the agro and animal health fine chemicals markets are similar, albeit less pronounced. Specialty chemicals outside life sciences, in contrast, are used by almost all industries and, therefore, virtually cannot be approached proactively. Also, the innovation rate in terms of new chemical entities is generally rather low, except in the electronic industry.

When looking after the "customer category/geography" matrix, it should be duly considered that the decision makers have their offices at the companies' headquarters. Under this premise, "pharma" the United States are the most attractive combination for fine chemical companies, followed closely by Europe.

> Many large pharmaceutical companies have established pharmaceutical manufacturing sites in tax havens such as Ireland, Puerto Rico, and Singapore. However, the decisions regarding the production programs and the procurement of, for example, PFCs are taken at the companies' headquarters. Under the perspective of logistics, however, these overseas locations are important destinations for the goods supplied under contracts concluded with the headquarters in many cases.

Within *ethical pharmaceuticals*, out of the top 10 pharma companies, 5 each are based in the United States and Europe, the former accounting for 43%, the latter for 57% of the $300 billion sales of ethical drugs (see Table 11.2). Although Europe, after the takeover of Genentech by Roche, is the frontrunner in terms of share of sales, the United States still remain the more interesting target market for fine chemical companies. The reason is that the Europeans traditionally are less inclined to outsourcing. As there are only a few domestic fine chemical companies in the United States, the market is also particularly attractive for non-USA-based fine chemical companies. In Europe, there are big differences among individual countries. The United Kingdom and Switzerland are most important. In England, both AstraZeneca and GlaxoSmithKline are large purchasers both for CM services and API-for-generics. GlaxoSmithKline, which has maintained also the former Smith Kline & French headquarters in Philadelphia, USA, has pioneered the concept of a "suppliers' day." The purpose of these one-day events is to align vendors with the procurement policy of the company. The statement of the head of Global Supply: "If you want to remain a supplier to our company, you must commit— in writing—to a 20% reduction of the price of every single product you sell us within the next three years, 7% per year to be precise" has left its mark in the fine chemical industry. AstraZeneca has announced its intention to divest all its chemical manufacturing plants. This is a completely different attitude as compared with Roche, which purchases only nonregulated intermediates for its new drugs! Also, Novartis maintains a strong in-house manufacturing base. Thanks to Sanofi-Aventis, France now holds the third place in the European pharmaceutical industry. No German company is left among the top 10—a strong demise after its global dominance for about a century. Actually, the successors of Germany's once-famous chemical giant IG Farben, BASF, Bayer, and Hoechst, have no longer been playing in the major league of the global pharmaceutical industry for many years. With the sale of Knoll AG to Abbott, BASF has exited pharmaceuticals completely. Bayer still successfully capitalized on the first synthetic drug ever invented, Aspirin. In the same context, it also had acquired Roche's over-the-counter (OTC) business, but has suffered serious setbacks with its proprietary drug business after the withdrawal of Baycol/Lipobay (Cerivastatin) in 2001. Hoechst's pharma business, such as those of Rhône-Poulenc and Sanofi, is now part of Sanofi-Aventis, and thus moved from Germany to France. After the acquisition of Ratiopharm by Teva, Germany's position on the generics market is also diminishing, despite the large home market. Japan ranks third with regard to both pharmaceuticals and agrochemicals. After the mergers of Fujisawa and Yamanouchi to form Astellas in 2005, and of Daiichi Pharmaceuticals and Sankyo to form Daiichi-Sankyo, also in 2005, there are three Japanese pharmaceutical companies among the top 20, the largest one still being Takeda.

Within *agrochemicals* there are three American, three European (German/ Swiss), and two Japanese companies among the top 10 (see Table 11.5). In terms of market share, the Europeans, Syngenta, BASF Agricultural Products & Nutrition, and Bayer CropScience clearly dominate with a 54% share of the top 10's sales of $41.8 billion, followed by Monsanto, Dow Agrosciences, and DuPont, 28%, and Sumitomo Chemical and Arysta LifeScience, 6%. The proportion of genetically modified seeds being smaller in the European agrochemical companies' portfolios as compared with their U.S. competitors, the "fine chemicals shopping basket" is even bigger than the percentage numbers indicate. Despite its small size, Switzerland holds a very prominent position in the global life science community. Four world-class life science companies are based in the small town of Basel (with 200,000 inhabitants), namely, the pharma rivals Novartis and Roche, and the world's number 1 agrochemical company, Syngenta, formed from the former pesticide divisions of Zeneca (respectively I.C.I.) and Ciba-Geigy, which is also the predecessor, jointly with Sandoz, of Novartis, and Lonza, the world's leading fine chemical company. Contrary to its peers, Roche originally was a pharmacy. The first successful product was a cough syrup, *Sirolin-Roche*, launched shortly after foundation of the company in 1896. Therefore, it has traditionally been more open to outsourcing than the industry average.

> An example of an enduring relationship between a pharma and a fine chemical company is the one Roche forged with the Italian fine chemical company F.I.S. It goes back to 1985. At that time, an F.I.S. salesman visited Roche trying to get some business. He was told by purchasing that Roche had well-established relationships with a number of companies and was not looking for a new vendor. The only chance to develop business would be if F.I.S. were in a position to supply Aditoprim, a veterinary version of the antibiotic trimethoprim at a lower price than Roche's internal manufacturing cost. The R&D department of F.I.S. evaluated the process and concluded that there was no chance to meet the price target.

At this point, the president-founder of F.I.S., Dott. G.-F. Ferrari, overruled the conclusion of the R&D department and decided that the business was to be done. Roche was so impressed with the performance that it concluded a 10-year frame manufacturing contract, under which F.I.S. custom-manufactured precursors of Roche's famous diazepine antidepressives. Dott. Gian-Franco Ferrari was always received with great respect for his entrepreneurship at Roche's headquarters at Grenzacherstrasse in Basel.

Roche was also one of the first companies to formulate an outsourcing strategy in the mid-1990s. The basic concept was to outsource APIs for mature drugs and use the freed capacity for new drugs, which were to be produced in-house. A tangible outcome of this strategy was the construction of a launch site in Florence, South Carolina, USA. It made history as the most expensive primary pharmaceutical manufacturing plant, on a $/m^3$ basis, ever built. After the acquisition of Syntex, USA, and Boehringer-Mannheim, Germany, Roche obtained access to large PFC production capacity. The first concern of the head of global manufacturing, which also included procurement, was to fill this capacity. Novartis is relying primarily on in-house production, recurring to third-party manufacturing only if technologies are required that Novartis does not want to perform in-house. An example in point is the blockbuster antihypertensive drug Diovan (valsartan). Novartis has expanded its in-house production capacity for the steps requiring only conventional chemistry, but is outsourcing an intermediate whose synthesis requires energetic sodium azide chemistry. Novartis was also relatively late in employing chemists in its purchasing department, a prerequisite for outsourcing. Also, the first supplier's day of the pharma division was held in 2010! Contrary to common belief, none of the Basel giants is giving a right of first refusal to Swiss fine chemical companies, such as Dottikon Exclusive Synthesis, Lonza, or Siegfried. The mere fact that the staffs of the Swiss life science giants are very international rules this out.

Japan, the world's second largest pharmaceutical market, has opened up to overseas fine chemical companies lately. On the one hand, laws forbidding imports of APIs have been shelved; on the other hand, Japanese pharmaceutical companies are approaching outsourcing with a more positive attitude. Also, the generics market is growing, albeit from a very small base. Offshoring the chemical manufacturing of PFCs for proprietary drugs is still the exception rather than the rule. It is a last resort only if the local fine chemical companies— which often are part of the same industrial conglomerate as the pharmaceutical companies themselves—are not in a position to satisfy the demand because of lack of either capacity or special technologies. Well-known cases of CM deals concluded with European fine chemical companies were involved in the supply of PFCs for Sankyo's Noscar, respectively Warner-Lambert's Rezulin (troglitazone), which, unfortunately, was withdrawn in 2000, and more recently to Takeda's blockbuster drugs Blopress (candesartan) and Prevacid (lansoprazol). Japan's number one pharma company, Takeda, has closed chemical manufacturing at its large Osaka site, and is relying more on outsourcing. It must be realized, however, that it takes many years to develop an important business relationship with a Japanese company. Only the large fine chemical companies have the means to sustain a year-long business development effort. Once established, the Japanese partner will honor its obligations beyond the letter of a supply contract.

The structure of the pharma industry (see Section 11.1 and Table 11.1) serves as basis for identifying which *customer categories* should be served. Each of them—big, medium, and small pharma—has advantages and disadvantages. In terms of potential business volume, "big pharma" ranks highest with a share of about 40% of the total PFC market (mainly "exclusives"), "medium pharma" ranks second with a share of about 30% (mainly API-for-generics), and "small pharma" comes last with about 6% (only "exclusives"). The building blocks cannot be assigned. On a "business potential per company" basis, "big pharma" obviously is most attractive (see Table 12.6). Also, target companies are easy to identify. However, competition for business is very strong, and those involved in procurement are well aware of the company's purchasing power. Only fine chemical companies that have a proven track record of superior performance and may have even achieved "preferred supplier" status have a realistic chance to be considered for new business.

Small or virtual pharma typically do not have established products and have only a limited new product pipeline. Therefore, they offer few business possibilities on an individual company basis. This is compensated by the large number of companies (well over 1000 in the United States alone) and their lack of manufacturing assets and expertise. Also, the number of approvals for new chemical entities (NCEs) originating from virtual pharma companies surpassed that of the top 20 for the first time in 2003. A substantial amount of deskwork is required in order to identify developmental drugs that have both

TABLE 12.6   Breakdown of the Merchant Fine Chemicals Market

Type of Company	Position on the Drug Life Cycle		Generic
	Patented		
	Phase II/III Access Point 2[a]	Phase IV Access Point 3[a]	Access Point 4[a]
"Big pharma"	$1.0 billion	$6.5 billion	$3.0 billion
"Medium pharma"	$<0.5 billion	$2.0 billion	$5.0 billion
"Small pharma"	$>0.5 billion	$0.5 billion	—
Subtotals	$2.0 billion	$9.0 billion	$8.0 billion
Building blocks	$4.0 billion[b]		
Grand Total	**$23 billion**		

[a] *See* Fig. 12.2.
[b] Do not require production under cGMP.

Source: Adapted from Prochemics, Zürich (April 2004).

a good chance of success and a fit with the technologies of a given fine chemical company. Also, its business is at risk once the small pharma company licenses its new drug to big pharma. The portion of new drugs licensed-in by big pharma increased from 15% to 60% of all new drugs between 1969 and 2009! The licensing agreement usually has a clause that gives the licensee full freedom to adopt its own purchasing policy. This includes also in-house manufacture. On the positive side, small pharma's share of the total business is growing above average, the competition for business is smaller, its procurement is less hard-boiled and, last but not least, depends totally on outsourcing.

Small or virtual companies are attractive for CROs. They require experienced specialists to assist them with various chemistry-related research services and are willing to sign on FTE agreements more easily than are big pharma companies. The sales cycle is short, and the headaches are minor in selling to virtual pharma. It has to be noted, however, that they prefer to deal with local companies.

Midsize (medium) pharma play a modest role in blockbuster drugs (all of the top 20 blockbuster drugs were marketed by big pharma companies in 2009!) but are important users of API-for-generics.

In conclusion, all three customer categories have advantages and disadvantages, and none should be excluded. The selection of the most appropriate distribution channels and key account management (see Sections 12.4 and 12.10, respectively) are useful tools for managing the universe of potential and existing customers.

As mentioned at the beginning of this Section, the United States remains the most important target *geographic region*. Apart from the large markets for proprietary and generic drugs alike, there is a trade deficit for PFCs. Within the United States, big pharma is concentrated on the East Coast, particularly the state of New Jersey. Out of the roughly 1200 U.S. virtual pharma companies, 500 are located at the West Coast. Europe is the second most important market. Both in terms of big pharma companies and generics penetration, it is closing up to the United States. Furthermore, a strategic shift to more outsourcing is taking place. Because the European big pharma companies are located in England (AstraZeneca, GlaxoSmithKline), Switzerland (Novartis, Roche), and France (Sanofi-Aventis), these countries are most important for developing a fine chemicals/CM business. The attractiveness for API-for-generics ranks by and large according to the population of the countries, which means, in decreasing order, Germany > UK > France > Italy > Spain > other countries. Although Japan, with a share of 11%, is the world's third largest pharma market, there are only three companies among the top 20. They hold a share of 10% of the top 20's global turnover. The corresponding numbers for Japanese agrochemical companies are 7 companies among the top 20, with a share of 10% of the global turnover. Japan is an arid place for non-Japanese fine chemical companies. Despite the fact that legislation protecting local pharmaceutical fine chemical manufacturers has been abolished recently, non-Japanese companies have a long way to go to get a significant stake in this traditionally protected market. A rapid decline of Japanese first launches from 19% in 1999 to scarcely 5% in 2004 [9] does not help, either. All in all, few examples of major supply agreements for PFCs between foreign fine chemical companies and Japanese pharma companies have been concluded. Pharmaceutical and agrochemical companies in Asia, Africa, and Australia have only a local reach. As more and more people can afford Western medicine in China and India, an above-average demand growth is expected in these highly populated countries. Because of a substantial and very competitive, albeit extremely fragmented, local production, the markets in these developing countries are practically inaccessible to Western fine chemical companies, unless they enter into joint ventures (see Chapter 14).

12.4 DISTRIBUTION CHANNELS

International commerce prevails in the fine chemical industry because suppliers and customers are often located in different countries, or even continents. Transportation costs are almost negligible. For managing their international business, fine chemical companies have to choose the most appropriate distribution channels. Basically, they can do it by their own means—either directly from their headquarters, or indirectly through a local office—or with the help of an agent or distributor. The total control of the supply chain is the main argument in favor of an in-house solution. Also, it avoids a conflict between the principal's goal of a long-term profit optimization and the agent's interest in short-term profit maximization. For this reason, agents generally are not interested in business development activities, which generate commissions after several years only. The main advantages of agents are their long-standing networks of customer contacts, and knowledge of the local legislation and business conditions. This is particularly important if customers and suppliers are part of different cultures, such as Eastern and Western cultures. Japanese trading houses (see Section 12.2.2) have an immense experience in bridging the gap. By representing different companies, agents can offer a large range of products and services. They are in a position to "open the doors" at accounts that otherwise would not be accessible, particularly for small companies with a limited product/services range. The disadvantages are potential conflicts of interest because of overlapping product ranges of principals and concerns regarding leakage of intellectual properties.

For the selection of the most suitable *distribution channel*, the following elements should be considered:

- Knowledge of the country of destination, entailing both familiarity with the culture in general and the extent of business experience in particular. If a European fine chemical company wants to enter the Chinese or Japanese market for the first time, the services of an agent are almost mandatory. On the other hand, if the same company has gained many years of experience selling its products and services to the United States through an agent, it will consider setting up its own office. When an agency agreement is prepared, care should be taken to formulate the "exit clause" in a way that does not unduly penalize the principal.

- Actual, respectively potential size of the business. If there is only a small sales potential in a given country, let us say in Taiwan, an agent is the distribution channel of choice. The minimum sales proceeds that are needed to justify a local representative, working out of his home, is about $5 million per year.
- Customer categories. Big pharma companies, which are mainly customers for exclusive products and have extensive procurement and logistics departments of their own, prefer to deal directly with the supplier's head-quarters, regardless of its location. Midsize and virtual pharma companies welcome the logistic assistance from agents, for example, for product registration, custom clearance, and local transportation.
- Product categories. Contract manufacturing projects, which require mul-tiline and multilevel interactions, are better managed through direct con-tacts; API-for-generics are preferably channeled through agencies, whose assistance for local registration is a valuable asset.

As not all elements coincide in actual business life, a compromise has to be made. In a frequently used approach, the key accounts are served directly from the headquarter (see Section 12.9), the other customers require contract man-ufacturing services by the local office, and the generics companies by special-ized agents as those mentioned in Section 12.2.2.

12.5 PRICING

Whereas controlling determines the cost of a product or service (see Section 7.2), marketing sets the sales price.

Three different methods are used for the pricing of fine chemicals (see Table 12.7). They are adopted depending on the development status (laboratory/

TABLE 12.7 Pricing Models

	Sample Preparation/ Laboratory Research	Industrial Scale Production
Exclusive products/ services	Time-based bottom-up	Volume-based bottom-up
Standard products (e.g., API-for-generics)	N/A	Volume-based top-down

industrial-scale) and on the business type (exclusive/standard). A *time-based* pricing is mainly used by CROs for samples produced on a small scale, or for process research. The most frequently used unit for invoicing is "$/year" of an "FTE" (full-time equivalent). A *volume-based* approach, either "bottom-up" or "top-down," is used for fine chemicals produced on industrial scale. The usual unit is "$/kg." The *time-based* approach factors in the time spent on the preparation, rather than the material cost of a sample, as the key cost element in the laboratory research phase. The FTE calculation is driven primarily by the remuneration of the researchers and varies considerably between developed and less developed countries (see Chapter 14). In order to allow customers to control the appropriate spending of the research work that they are funding, "milestone payments" are mutually agreed upon. They condition FTE payments on the fulfillment of certain pre-established objectives, for instance, yield, throughput, or quality targets. Another pricing model for CROs gaining popularity is "risk sharing" by which the CROs earn milestone payments when they complete certain pre-agreed research objectives.

In the *volume-based* model, a "bottom-up" calculation is used in exclusive products, where there is no established market price, as opposed to a "top-down" calculation in the case of API-for-generics. In the former case, the target price is calculated on the basis of the raw material and the conversion costs—which individually account for about 35% and 55%, respectively, of the sales price and together represent the COGS—and a profit margin is added. As the raw materials are the single most important cost element, particularly if the synthesis starts at a late stage, they require particular attention. Issues that have to be addressed are "how the consumption can be reduced," "which are the minimum specifications," "make or buy," and, in the case of "make," whether the internal COGS or the market price should be used for the comparative calculation. Both in the "bottom-up" and "top-down" variety, the main factors are the same, but they are applied in opposite sequences. Thus, if the subtraction of profit, conversion, and raw materials cost from the market price for an API-for-generic results in a negative number, either there must be means to reduce single cost factors, or the business is not viable. Depending on the overall market situation in general, and specific customer–supplier relationships in particular, the axiom that prices must be higher than costs has to be violated and a reduced profit or even a loss have to be accepted in some cases. A number of "freebies," respectively services without an immediate return, are part of the market strategy. In order of increasing expense, they range from free offers, free samples, free pilot plant tests, pilot plant quantities at "industrial-scale" prices, execution of non-qualifying projects, capital investment in special equipment, plant adaptations, and capacity expansion all the way to new plant constructions. When setting a price, one must be aware of the fact that the specialty chemicals industry makes formulated products, of which the active ingredients represent only a fraction of the COGS. This is particularly the case for the pharmaceutical industry. It considers the active ingredients of the drugs, aka the drug substance as raw materials. They account

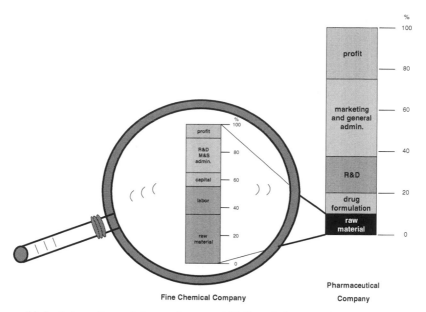

Figure 12.4   Integration of drug substance (API) and drug product price structures.
Source: European Association of Generic Producers, IMS Health Service, CGEY, 2002.

for less than a 10% fraction of the sales price. In other words, the drug that is sold in the drugstore is at least 10 times more expensive than the drug *substance* (API), which constitutes the raw material for the pharma and the finished product for the fine chemical industry. The price of the drug *product* is composed of four main elements: COGS (drug substance, formulation + packaging), ~20%; R&D, 15–20%; marketing and general administrative expense, 35–40%; and profit, ~25% (see Fig. 12.4). Therefore, some flexibility regarding the expected cost, which is the result of the subtraction of four not exactly defined numbers, can be expected. Actually, the pharma industry's internal "top-down" calculation base for the pricing of a given drug product is primarily driven by the patient, respectively the daily medical cost of the treatment, expressed in $/*day*, within a given therapeutic category of comparable products on the market. Subsidiary elements are the patent status, the strength, the frequency of administration, the therapeutic category (it is higher for anticancer or anti-AIDS drugs than for more traditional categories, such as cardiovascular and CNS drugs), and the country in which the drug is sold (see Fig. 12.4). Neither the complexity of the molecule nor the dosages are determining factors. Also, the prices for different strengths of the same drug differ only marginally in most cases. For instance, a package of thirty 50-mg tablets costs only a few percentage points more than one with the same number of 25-mg tablets. This is another indication of the small relevance of the API unit cost. Minor price variations may occur on the basis of specific advantages in terms of the risk profile, the convenience of administration, the image of the brand, and other variables.

The appreciation of the fact that the pharmaceutical industry uses a completely different cost calculation for the drug substance is an essential prerequisite for successful price discussions. For high-dosage drugs made of complex molecules, the different approaches can lead to big spreads with regard to the drug industry's expected target cost on the one hand and the sales price expectation of the fine chemical company on the other hand.

> A comparison of the daily medical costs for the drugs made for antihypertensive drugs, illustrates the pharma industry's approach. Regardless of the sales revenues of the drugs, which varied between $4.7 billion for Pfizer's Norvasc and $30 million for Solvay Pharma's Teveten, the medical cost was in the range of $1–$1.40/day for the low dose—typically 20 mg—and $1.50–$2/day for the high dosage. Within these ranges, the oldest subcategory reported, namely, β-blockers, introduced in the 1970s, were the cheapest at $1/day (at the low-dose level), followed by ACE (angiotensin-converting enzyme) inhibitors, $1.05/day; calcium antagonists, $1.28/day; and finally the latest subcategory, the angiotensin II receptor antagonists (or simply "sartans"), $1.40/day.

This example discredits the conventional wisdom that drug prices are open-ended. At least for "we too" drugs, there is competitive pressure within the same therapeutic category. Awareness of the "daily medical cost" is all the more important for the supplier, as typically there are no reference prices that can be used as a guideline for checking the figure with which the internal controlling has come up. Furthermore, the confidentiality agreements, which exclude any information exchange with third parties, prevent suppliers from applying traditional marketing principles in order to determine whether their prices are competitive. At the end of the day, the likely winner of the impasse is the innovative fine chemical company, which invents a breakthrough process enabling drastically reduced COGS. From pharma's perspective, neither the complexity of the molecule nor the dosages are determining factors. Actually, the prices of different strengths of the same drug differ at most marginally in most cases. Thus, at the Internet pharmacy drugstore, 100 tablets of Lipitor cost $419.95, the same for 20-mg, 40-mg, and 80-mg tablets! This is another indication of the small relevance of the API unit cost. Minor price variations may occur on the basis of specific advantages in terms of the risk profile, the convenience of administration, the image of the brand, and other variables.

The attainable profit margin depends on both customer- and supplier-driven factors. The former includes the competitive situation for the final drug (therapeutic category, innovation, dosage, risk/reward profile, etc.); the latter, the competitive intensity for the product sold. Last but not least, the dollar value of the business and the overall business condition should be considered. Given all these elements, the profit element can vary substantially. In the worst-case scenario, not even all fixed and overhead costs can be recovered; in the best-case situation, the markup can be up to 100% or more, if a supplier is in a particularly favorable situation. In order to narrow this uncomfortably wide range, a "value-pricing" approach can be helpful. It is used frequently in the specialty chemicals industry and consists in analyzing the value that a product or service presents to the customer. Questions that should be addressed in this context are

- What is the share of the price of your product as compared with the end product?
- Does the quality of your product allow the customer to differentiate from the competition?
- Are you the only supplier who can guarantee on-time delivery and sufficient default production capacity?

In the case of a tolling contract, where the customer supplies raw materials "free of charge," the markup for profit is calculated as if the material were purchased, provided that the material cost does not exceed the conversion cost.

In Figure 12.5 the price structures for a PFC and for a pharmaceutical are combined. Particularly noteworthy are the differences between the cost items "R&D" and "marketing and general administrative expense" (~50% for a finished drug; 25% for a PFC), and "raw materials" (10% vs. 35%).

A price calculation is relatively easy for a product with a track record of a regular industrial-scale production. On the other hand, it is difficult, if only a

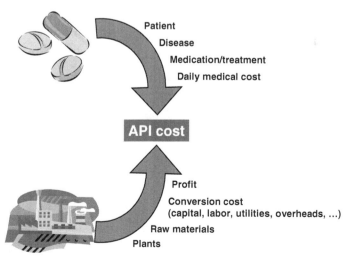

Figure 12.5 Pricing mechanisms of the pharmaceutical and fine chemical industries.

laboratory procedure exists and a calculation has to be made based on the virtual scale-up to industrial-scale production, especially if you have to rely on incomplete information from your customer. Typical pitfalls are

- A quote for 5000 kg is requested, but the target product actually had never been made on more than a gram scale.
- Each step of chemistry was performed on a different scale ("mosaic chemistry").
- Only the synthesis for a distant analogue is shown.
- Procedure conveyed is eerily close to the published patent literature, only at a 100 times larger scale ("paper scale-ups").

The ability to perform this desk exercise in a quick and reliable fashion is an important competence criterion of a fine chemical manufacturer. If mistakes in the cost or price calculation have been made and prices have to be changed, you will need facts to support your request. If a pivotal product is supplied, a supply contract is concluded.

A serious threat for Western fine chemical companies is low quotes from competitors in Asia. In this context, it is worthwhile to draw the customer's attention to the fact that the "ex works" unit price of a product is only part of the story. For a meaningful comparison, it is mandatory to determine the *total cost of ownership*. Apart from the "ex works" product price, it comprises the sales terms and delivery conditions (including packaging, transportation, and custom duties), the travel expenses in connection with visits to the supplier's premises, costs for auditing and qualifying the Asian supplier and for registering the product. Furthermore, abnormal situations, such as delayed or incomplete deliveries, handling of quality complaints and IP violations have to be taken in due consideration, too. Thus, the total cost of ownership of the purchase can be much higher than the invoiced price—and make the Asian option unattractive. As some of the aforementioned expenses (they may also include the cost of running a procurement office in Asia) are fixed costs, the Asian solution is the less attractive, the smaller the size of the deal. It is also unappealing in the case of the need for an urgent delivery, or an exclusive product as compared with a standard product. In the latter case, IP issues and dependency on one supplier are not at stake. The issue boils down to a trade-off between a $30 million revenue loss, if, say, a $1 billion per year blockbuster drug runs out of stock for 10 days because of delayed receipt of the API against a few percentage points lower unit cost. ...

An example in case is a leading German pharmaceutical company, which had chosen two Chinese fine chemicals companies for the supply of a pivotal intermediate. When interruptions of shipments occurred, it found out that the starting material for the intermediate originated from just one source, which local authorities had closed down after serious environmental problems had arisen.

12.6 INTELLECTUAL PROPERTY RIGHTS

IPR are *the* most sensitive interface between a fine chemical company and its customer. This is particularly the case in CM deals with a life science company. Most of its profits derive from human drugs, veterinary drugs, or agrochemicals protected by product patents. Any dissipation or misuse of IP, on either the product or the manufacturing process, can cause serious damage. The company, its board, executives, and employees may be held liable. It is, therefore, imperative that strict procedures for safeguarding the IP are put in place, such as the following:

1. A one-way nondisclosure agreement (confidentiality disclosure agreement [CDA]) has to be signed by board members, management, employees, and external consultants. The individuals are bound to the CDA also in the case that they leave the company (see also item 10 below).
2. Employees susceptible to work with confidential information must be trained regularly. For individual tasks, they must sign a specific CDA.
3. Documents that contain confidential information (e.g., laboratory journals, batch records, campaign reports), have to be earmarked and signed at regular intervals by the project manager. The latter surveys the circulation and copying.
4. Sensitive information has to be saved exclusively in these earmarked documents, which are maintained in a secure place and remain within the premises at all times.
5. All documents and samples must be maintained for at least 5 years.
6. The access to confidential data must be restricted for each employee according to his work requirement. Thus, the bench chemist is only authorized to retrieve information on the part of the synthesis he is working on, the group leader on the whole synthesis, the project manager on the whole synthesis, the marketing data, and so on. Using IT, this can be done, for example, by storing all documents in a suitably programmed server.
7. Customers stay updated on project progress through secure Internet sites, video, and teleconferencing.
8. Customers' names, projects, and compounds must be coded so that only management knows these details. The laboratory, pilot plant, and industrial-scale plant chemists and engineers know the codes and refer to each project accordingly.
9. Customers must be immediately notified of any breach in confidentiality.
10. Particular attention must be paid to employees leaving the company. They must not carry any written documentation with them.

Whereas IPR on *product* patents unequivocally belong to the customer, *process* patents arising from the cooperation between customer and supplier on a joint project are a controversial issue. The customer holds the position that all IPR resulting from a joint project belong to him. His arguments are (1) that without his input there would have been no invention anyhow and (2) that he has paid for the development work, and therefore the resulting IP, either directly by funding the supplier's R&D work or indirectly through the price that he pays for the product. The supplier maintains that IP is an essential prerequisite for conducting her business. If she cannot advance her know-how, she will drop out of business sooner or later. Also, the customer would not have entrusted her with a project, if he had not been attracted by her know-how. These opposite views are a critical element in contract negotiations and can become a deal breaker. A pragmatic approach to resolve the problem is to allow the supplier to use the IP outside the area of the customer's direct interest. In the narrowest interpretation "direct interest" means that she is authorized to use the process IP for syntheses of fine chemicals outside the product that is the object of the joint project. If the product is a drug, a more obliging interpretation could be that the product has to be outside the specific therapeutic category, for example, "sartans" in the case of antihypertensives. In agrochemicals, "pyridine fungicides" could be a compromise. As a last resort, usage could be restricted to non-pharmaceutical, respectively non-agrochemical purposes only.

For reflections on the IPR situation in Asia, refer to Section 14.2.

12.7 SUPPLY CONTRACTS

A supply contract is the most commonly used document for validating a business deal between buyer and vendor, particularly in CM. They even more and more substitute—a.o. for regulatory requirements—simpler purchase orders used for standard products' one-shot deals. More extended cooperation agreements, such as joint ventures, are discussed in Section 14.3. Over and above the more and more elaborate legalistic wording of the contracts, one should always bear in mind that the three main elements for a successful business partnership are *cultural fit, strategic fit*, and *operational fit*.

> Novartis emphasizes finding the right partner, rather than doing the right deal. If the partner is right, the deal will succeed. The emphasis is also on effective communication from the get-go. Decide what the vision is for the collaboration, and make sure there is buy-in by both parties. Identify responsibilities for each partner and set up methods for resolving differences. Share information freely. Evaluate performance just as you would for an internal project. Be honest. Talk freely and often. Make decisions based on facts. Invest resources in the success of the partnership.
>
> —Novartis deal-making policy

On an industrial level, supply contracts are product-based (unit price in $/kg), whereas in contract research agreements, they are service-based (FTEs in $/scientist-year, see previous Section). The two types of contract are discussed in more detail hereafter. Agreeing on the price for a fine chemical is only one, albeit essential, element of a *CM* deal between the supplier (fine chemical company) and the customer (specialty chemical company). The supply contract entails a considerable financial exposure (sometimes hundreds of millions of dollars), covers an extended period of time (typically 3–10 years), and must factor in a number of imponderable elements of the cooperation. It is mandatory, therefore, to define the obligations of the partners and to hedge against unpredictable events. Prior to entering into the contract negotiations, the partners should clearly define the scope and objectives that they wish to achieve. It should have the necessary provisions to cope with "what if's," such as delays in drug approval/start-up of production, substantial increase or decrease in demand (in the worst-case scenario, withdrawal of the drug), failure to meet the agreed-upon yield and throughput figures, unsolicited offers from third parties, subcontracting, takeover of one partner by a competitor, exchange rate fluctuations, force majeure incidents, and so on. The main elements of a contract are the "commercial clauses," that is, quantities, prices, and supply logistics; the "technical clauses," that is, specifications or description of the services the custom manufacturer has to provide, including milestone plan; and the "legal clauses," that is, IP ownership (see Section 12.6), warranties, indemnities, exit, and other boilerplate clauses that lay out the parties' reciprocal responsibilities. The commercial and technical clauses are drafted by specialists from the involved activities. Key commercial, technical, and legal elements of a supply contract are listed in Table 12.8.

For the sake of good order, the contract can be subdivided into four phases: Phase I relates to the planning and configuration of the partnership. Phase II

TABLE 12.8   Key Elements of a Supply Contract

Commercial	Technical and Regulatory	Legal
• Product	• Product specifications	• Duration (extension/
• Quantities	• Process description	cancellation)
• Prices (price/qty/	• Analytical methods	• Investment guarantees
third party offers)	• Process improvements	• Force majeure
• Currencies	• Change control	• Insurance coverage
• Forecasts	• Plant description	• Confidentiality
• Provision of	• Quality control/quality	• Intellectual Property
starting materials	assurance	Rights
• Call-off orders	• Batch records	• Liabilities[a]
• Shipments	• Audits and inspections	• Compliance with laws and
• Packaging/labeling	• DMF (drug master file)	regulations
• Back-up capacity	• Safety, health, and	• Applicable law/arbitration
	environment	• Exit strategy

[a] Liability for consequential damage, such as claims of patients against the drug company, are usually waived.

comprises the period from the project initiation to the introduction, resp. verification and validation of the system. Phase III comprises the actual execution of the contract manufacture. Phase IV covers the exit plan. If litigations arise, arbitration rather than court litigation is the usual way of settlement. As a consequence of the deteriorating business condition for the fine chemical industry, the bargaining power of the customers has increased, and the contract terms have worsened for the suppliers (see Table 12.9). As part of a supply agreement, customers often ask for "cost transparency." This is a biased request, which can cause unpleasant discussions on the *just* costs of goods sold. The key word that pops up most frequently is *cost allocation*. Typical problems are the fixed cost allocation, for instance, the cost of nonutilized parts of a production train or a plant, or the allocation of overhead costs.

Frequently asked questions by purchasing are "Why do you include an allocation for R&D costs, when we already have funded your R&D expenses through a milestones payment agreement?" or "Why do you charge 5% for marketing and sales, when all what you need to do is to ask your commercial assistant to call off a truckload once a month?"

For *contract research* contracts, two cooperation modes are used:

1. *Full-Time Equivalents (FTEs).* A defined number of scientists are assigned full-time to a customer's project. Whereas the CRO provides the infrastructure, the tasks are given and the work is supervised by the customer. The FTE charge is inclusive of the scientist's salary, raw materials, and other consumables (up to an agreed-on maximum), attending milestone meetings at the customer's site, and analytical and shipping

TABLE 12.9   From a Seller's to a Buyer's Market

	Prior to Year 2000	After Year 2000
Contract duration	5 years	1 year
Capital guarantee	Yes	No
Take or pay clause	Yes	*What is this?*
Number of suppliers	Sole	Several
Supplier loyalty	Partnership	"Auction"
Volume forecasts	Binding	Spot orders
Price adaptation	Upward: price index, etc.	Downward: ×% yearly reduction
Process improvements/ customer inventions	To supplier	To customer
Penalties for off-take delays	To supplier	To customer
Participation at plant Adaptation costs	Yes (for specific investments)	Case-by-case
R&D expenditures	To supplier	To customer

cost. Indicative numbers are \$150,000–\$250,000 for United States and Western European countries, around half this figure for former Eastern European countries, and \$40,000–\$60,000/year for China and India.

The key elements of an FTE agreement for "chemistry type" work are

1.1 Obligations of the CRO
- Perform synthetic chemical research, process development, and optimization for any projects entered in
- Provide technical consultation and assistance
- Utilize existing and develop new analytical methods
- Prepare samples, together with certificates of analysis
- Provide progress reports at regular intervals and a final report
- Provide all experimental records and laboratory notebooks

1.2 Obligations of the customer
- Provide appropriate technical assistance
- Organize project review meetings at regular intervals
- Fulfill its financial obligations

1.3 Confidentiality as per separate CDA

1.4 Payment terms and invoicing

1.5 Duration and cancellation

1.6 Arbitration

2. *Time and Materials.* This is more of a consulting business model. The customer is being charged for the total time spent by the CRO on a particular project. Material used, travel expenses, and other out-of-pocket expenses are also billed to the client.

12.8 PROMOTION

Promotion is an indispensable marketing tool in the fine chemical industry. It extends all the way from personal contacts with existing and prospective customers, customer events, booths at trade fairs, distribution of company brochures, advertisements in traditional media and the Internet, and websites. Whereas personal contacts are most likely to generate business, nonspecific promotional materials cultivate the "brand recognition" of a fine chemical company. Three particularities have to be duly considered: (1) as the business transactions occur within the chemical industry, specialized—and not mass—media have to be chosen for conveying the message; (2) the small size of most companies calls for a careful budget management; and (3) in contrast to advertisements of the consumer goods industry, a direct generation of sales cannot be expected. The primary scope of promotion is to cultivate the "brand recognition" of a company. It should give customers the confidence that they have made the right choice. In Table 12.10, a number of promotional tools are listed

TABLE 12.10 Cost-Effectiveness of Promotional Tools

		Cost		
		High	←	Low
Efficiency ↑	High	• Booths at trade fairs	• Customer visits • Plant tours • Customer events	• Networking/key acc't mgmt. • Articles in trade magazines • Interviews with executives
	Low	• Advertising in trade journals • Advertising in daily newspapers	• Poster sessions • Company brochures • IT webinars • Monthly "News from …"	• Company websites • Press/news releases • Internet advertising • Product listings

in a cost–reward matrix. The underlying consideration was the capability to attract and keep customers. For other objectives, such as attraction of investors or young talents, the ranking would be different.

Under these premises, key account management, cultivating personal contacts at all hierarchical levels, publishing articles in trade magazines, and having top executives interviewed by editors of well-known news media are the most cost-effective promotional tools. Personal contacts with existing and potential customers, either as visits to their offices or as tours at the suppliers' plants, are equally efficient, but are more costly. The participation at the leading trade shows, namely, CPhI (Chemical and Pharmaceutical Ingredients), Informex, and Chemspec, are expensive, but they allow a large number of contacts. It is recommended to arrange appointments ahead of time and also to take advantage of the presence of many customers for organizing an evening event. Websites have become useful in providing firsthand information. In order to be effective, they must be updated regularly and be accessible with a few mouse clicks. The latter is a challenge for large chemical companies, which have a small fine chemical business unit only. Advertising is a somewhat controversial promotional tool. On the one hand, a customer will hardly place an order on the basis of an ad; on the other hand, it assures existing customers that they have chosen a valuable supplier. Rather than generating inquiries from new customers, the advertiser will receive a lot of mail from agents looking for an additional principal. The magazines with the largest target groups and thus the first choice for advertising are *Chemical & Engineering News*, *Chemical Week*, *Scrip*, *ICIS Chemical Business*, and *Specialty Chemicals Magazine*. Company newsletters are the least efficient promotional tool. Recipients will receive dozens of monthly newsletters from all kinds of stakeholders, which usually are discarded immediately.

The *contents* of the promotional material are provided by business and corporate development. As the case may be, they will seek assistance from technical functions. The *form* of promotional literature and the *organization* of events are the task of the advertising department in a larger company, respectively of specialized agencies for smaller ones. There is no room for amateurism here any more!

The budget for the out-of-pocket expenses for promotion typically represents not more than 0.5% of sales.

12.9 NETWORK AND CONTACT DEVELOPMENT

The network of a fine chemical company constitutes a very valuable intangible asset. As it is the sum of the individual networks of its employees, a conservative company with long-tenure employees, who are prepared to share their know-how with their peers, is definitely in an advantageous position. Apart from enabling direct contacts between proficient business partners, it is of great help for resolving problems outside the daily routine and is thus a powerful facilitator for the future development of the company. It extends from issues such as filling open positions with qualified persons, improving competitor intelligence, obtaining authorizations or tax incentives from local authorities, to resolving legal questions in the United States, or establishing a foothold in Asia. In the context of the supplier–customer interface, it helps to get a better understanding of the customers' needs by opening information channels outside the normal business contacts in general and gaining knowledge of the importance of a particular project in particular.

Face-to-face customer contacts have been indicated above as the most efficient tool for the safeguarding of existing and acquisition of new business. Customer visits are a three-step effort, namely, preparative deskwork, the visit itself, and the follow-up, which all need careful attention. For details, see Appendix A.8, Checklist for Customer Visit. A tailored 20–30 min company slide presentation is the centerpiece of the visit. An example for the contents is given in Appendix A.9, Outline for a Company Presentation.

Depending on the type (standard vs. exclusive products) and the importance of a business transaction (small single-order vs. major multiyear supply contract), the *levels of customer intimacy* can vary between purchasing through the Internet with no personal contacts between the parties involved on the one hand and frequent face-to-face meetings between the commercial (and other pertinent) functions of the involved companies on the other hand (see Fig. 12.6).

As a rule, the more important the deal, the closer the customer intimacy. If a research laboratory buys a 100-g flask of sodium bicarbonate from a laboratory reagent supplier, it will place the order through e-mail. If a virtual pharma company entrusts the total production of the API for its only drug candidate to a custom manufacturer, a strong personal relationship is mandatory.

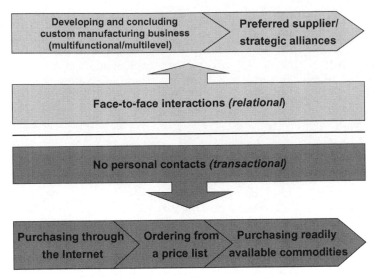

Figure 12.6   Levels of customer intimacy.

Customer–supplier contacts are facilitated if their respective organizations mirror each other. Various functions of the companies have different frequencies and intensities of contact. They are most intense between the commercial functions. Whereas only these counterparts should discuss business aspects of a project, other functions, such as R&D, logistics, back office, production, legal, and regulatory, are encouraged to have direct contacts, too, regarding specific issues of their particular competence.

12.10   KEY ACCOUNT MANAGEMENT

In order to manage the whole universe of customers, a differentiated management according to their relative importance is advisable. In Figure 12.7, it is represented as a pyramid, which is subdivided into three tiers. Tier I comprises those three to five companies that are considered as *preferred* customers, primarily on the basis of the importance of their present and future business. Tier I customers generate approximately 75–80% of total sales. They benefit from key account management and are prime candidates for a collaborative cooperation, as explained below. Approximately 20 *current and potential* customers make up Tier II. With them, there is either regular, albeit not spectacular, business or there are good prospects for important future business. Companies with occasional small deals are grouped in Tier III, *minor and peripheral*. They constitute the largest number but the smallest share of total sales. Tier I customers are accommodated by the headquarter (see Section 12.4, Distribution Channels); Tier I and Tier II customers are visited regularly

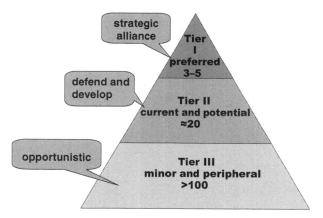

Figure 12.7 Key account management.

according to a standard annual program; Tier III customers are visited on an opportunistic basis only. The customer mailing list has to be set up in a way to allow a differentiation according to this classification.

An example of a collaborative partnership is BASF's effort to identify new sources of value for its automotive OEM (original equipment manufacturer) customers, which ultimately led it to install and run their paint shops as an integral part of their customers' assembly lines—a far cry from just supplying paint.

The prerogatives for a successful collaborative relationship, even more so than just for a supply contract, are a cultural, strategic, and operational fit between the partners. It also must be their explicit will to assign experts to joint teams that meet regularly, identify opportunities for synergistic business solutions, and develop and implement programs for their execution. The success of this relationship also depends on a highly skilled collaboration manager, who has a deep understanding of the partners' business and can break down internal and external barriers, align the various players, and drive out waste. He has to meet the multiple challenge of aligning the participating companies both individually and collectively across geographies, product families, and business units. As collaborations that are profitable as a whole can adversely affect the balance sheets of individual business units, appropriate metrics have to be developed to adequately compensate the persons concerned. In the abovementioned example of BASF's paint shop, the business unit "automotive paints" could lose from the new concept, depending on the transfer price for the paint.

It is not uncommon that personal friendships develop between individuals, who have taken an active role in shaping a strategic alliance.

BIBLIOGRAPHY

CITED PUBLICATIONS

1. D. Aboody and B. Lev, *R&D Productivity in the Chemical Industry*, New York University, Stern School of Business, New York, 2001.
2. B. Lev, R&D and capital markets, *Journal of Applied Corporate Finance* 11 (**4**), 21–35 (Winter 1999).
3. M. Angell, *The Truth About the Drug Companies*, Random House, New York, 2004.
4. Chemical Pharmaceutical Generic Association, *The Apis World Market*, Milan, Italy, 2009.
5. M. Bloch et al., Pharma leaps offshore, *McKinsey Quarterly Newsletter*, July 2006.

FURTHER READING

P. Pollak and E. Habegger, Fine chemicals, in *Kirk-Othmer Encyclopedia of Chemical Technology*, 4th ed., Wiley, New York, 2004.

The Merck Index, 14th ed., Merck & Co. Inc, Whitehouse Station, NJ, 2006.

D. S. Tomlin, ed., *The Pesticide Manual*, 14th ed., BCPC, Alton, Hampshire, UK, 2006.

PART III

OUTLOOK

CHAPTER 13

General Trends and Growth Drivers

The future of the fine chemical industry as a whole depends mainly on the destiny of the life science sector, in particular the pharmaceutical industry, and its R&D and manufacturing outsourcing strategies. While the pharma industry will remain the major customer segment, its demand for pharmaceutical fine chemicals (PFCs) is becoming more volatile and therefore less predictable. This is not so much the case for biopharmaceuticals (see Chapter 15) and active pharmaceutical ingredients (see Chapter 17), where globally a low double-digit growth can be expected, but it is very much the case for custom manufacturing of PFCs for proprietary drugs. This blurred outlook is due primarily to the uncertainties surrounding the impact of the health-care reforms in the United States and other countries and of the ongoing reviews of the business model by big pharma companies on innovation and procurement (see Chapter 16). The strategy reviews are caused mainly by the forthcoming revenue losses due to the so-called "patent cliff." Most companies will be unable to compensate them by sales generated from new drugs. These factors will determine the future demand for both contract research and custom manufacturing services. Depending on whether one takes the optimistic or the pessimistic view, either an increase or a decrease in demand is possible (see Table 13.1).

In addition to changes in demand patterns, ongoing globalization will have an important and probably irreversible effect on the location of the centers of gravity of the fine chemical industry. If the Western top-tier companies do not succeed in maintaining their "preferred supplier" status with the—still EU- and US-dominated—life science industry, the beneficiaries will be the emerging Asian fine chemical companies. The latter will also be the first ones to benefit from the rapidly growing growing domestic demand. As will be shown in Chapter 20, Western companies must have their own subsidiaries in the Asian markets if they want to take advantage of the business opportunities.

As a surrogate to presenting discrete growth figures for the industry, which would be presumptuous given the wide gap between positive and negative

Fine Chemicals: The Industry and the Business, Second Edition, by Peter Pollak
Copyright © 2011 John Wiley & Sons, Inc.

TABLE 13.1   Key Growth Drivers: Optimistic/Pessimistic Outlook

	Optimistic		Pessimistic
New drug launches	The new R&D structures and technologies adopted by pharma come to fruition and reverse the slump in new product launches.	↔	Pharma does not succeed in "collecting the high hanging fruit" and more stringent rules (including economic ones) further hamper new product launches.
Outsourcing policy	Pharma abandons chemical manufacture of APIs in favor of outsourcing.	↔	Pharma recurs to outsourcing only as last resort in case of capacity constraints or hazardous chemistry.

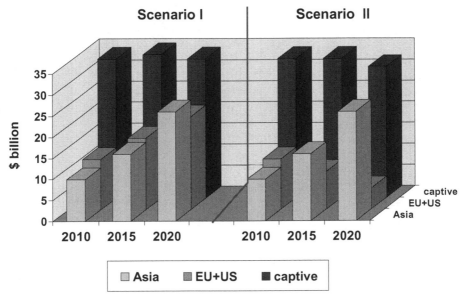

Figure 13.1   API growth scenarios.

growth drivers, two scenarios are presented for the further development of the most important segment of the fine chemicals market, APIs (Fig. 13.1).

The following assumptions have been made:

1. The global production value for APIs (captive + merchant) is $55 billion in 2010 (see Table 9.3).
2. The pharmaceutical industry—and as a direct result, its demand for APIs—will grow at a rate of 4% per year (scenario I) or 1% per year

(scenario II) between 2010 and 2020. In scenario I, the sales lost due to the "patent cliff" are overcompensated by the sales gained from new product launches and the growth of demand in the emerging markets. In scenario II there is no growth. Sales from new products are counterbalanced by sales losses due to patent expirations, and sales increases in emerging countries are offset by the negative impact of healthcare reforms in developed countries.

3. There is a 60/40, or $33 billion/$22 billion ratio of captive versus third party production in base year 2010. Due to ongoing manufacturing restructuring in the pharma industry, the share of outsourcing increases 4% per year in scenario I, and 2% per year in scenario II, to reach 60% and 50% respectively in 2020.

4. In comparison with the past five years. Asia's merchant API production value will increase at a somewhat reduced rate, namely 10% per year in both scenarios.

Three main conclusions are:

- Due to the impact of reduced growth of the pharma industry on the one hand and increased outsourcing on the other hand, the value of captive API production will remain flat at about $33 billion throughout the 2010 to 2020 period.
- In scenario I, the API production in the European and U.S. fine chemical industries will continue to grow, albeit at a slower rate than Asia's (+6% p.a. vs. +10% p.a.). In scenario II, EU + US will be negatively affected by the double impact of flat demand and globalization. Production value will decrease from $12 billion to $6 billion. In both scenarios, the production value of the Asian fine chemical industry will increase from $10 billion to $26 billion in the period under review.

The trends will be discussed in detail in the following chapters. The focus will be both on exogenous factors (i.e., demand from the pharmaceutical and agrochemical industries) on the one hand, and endogenous factors (globalization and biotechnology) on the other.

CHAPTER 14

Globalization

The fine chemical industry, like others, is subject to the global shift of manufacturing and service activities from high-cost, developed regions to low-cost,
emerging regions of the world. Next to the developments on the demand
side, globalization is the trend with the largest impact on the fine chemical
industry.

From a historical perspective, the increasing importance of Asia should not
come as a surprise. The Eastern hemisphere, especially China and India
(Chindia, as it is often called today), has held superpower status throughout
most of the history of humankind. The twentieth and early twenty-first
centuries were exceptions. Asia had an almost 40% share of the world GDP
in 1870. It went through a trough in the last century, reaching its lowest level,
19% in 1950, but has recovered to 25% in 2008. With a share of 7%, China
now ranks third in the world. The corresponding figures for India are 2% and
twelfth. As Asia is becoming the "workbench of the world," an increasing
number of fine chemical research institutes and plants are emerging in this
region. The major part of them are built, operated, and owned by domestic
companies. Whereas the United States and Europe are losing ground, Asia,
especially Chindia, with its "high-skill/low-cost" advantage, is the main beneficiary. That is why the slogan has been extended to "high skill/low cost/high
future." "High-skill" manpower is abundant; close to 90% of all scientific PhDs
are Asians.

Labor costs in China and India are $4000–$5000 per employee per year as
compared with $50,000–$60,000 per year in the United States and Europe. The
labor cost advantage is somewhat mitigated—though by now means offset—
by lower labor productivity of $65,000–$70,000 per employee per year in
Chindia, as compared with $250,000–$300,000, respectively, in the United
States and the EU. Combining both sets of numbers, however, the ratio is still
close to 2.5:1 in favor of Asia (see Table 14.1).

For an estimate of the price structures of fine chemicals in the two regions
(see Table 14.2) based on these numbers, it is assumed that

Fine Chemicals: The Industry and the Business, Second Edition, by Peter Pollak
Copyright © 2011 John Wiley & Sons, Inc.

TABLE 14.1 Fine Chemical Industry Labor Cost and Productivity in Key World Regions

Country	Labor Cost ($000/ operator/year)	Labor Productivity ($000/employee/year)	Ratio	
			abs.	US = 1
United States	50	250	5	1
Western Europe	60	300	5	1
Eastern Europe[a]	16	126	8	1.6
India	5	70	14	2.8
China	4	65	16	3.2

[a] Hungary.

TABLE 14.2 Price Structure for Pharmaceutical Fine Chemicals in India and the United States

Cost Elements	Detail	United States	India
Raw materials	Raw materials, solvents, catalysts, utilities	**35%**	**45%**
Conversion cost	Direct labor	10%	9%
	Capital cost (depreciation + interests)	13%	6%
	Overhead (maintenance, R&D, QA/ QC, SHE, M&S, general administration)	25% **55%**	15% **30%**
Profit	EBITDA margin	**15%**	**25%**
Total	Ex-manufacturer's sales price	**100%**	**100%**

• The cost for raw materials is somewhat higher, and the cost of utilities is markedly higher in Chindia than in the EU and the United States.
• The productivity-adjusted labor cost in India is 40% of Europe's.
• Capital cost and overheads in India are 50% of Europe's.

By any financial measures, such as labor cost, investment cost, and manufacturing cost per cubic meter per hour, companies in Chindia fare much better. Table 14.3 shows a comparison of the operating costs for running current Good Manufacturing Practice (cGMP) fine chemical plants in terms of location, reactor size, and ownership. A plant equipped with standardized $6.3\,m^3$ reaction vessels is taken as an example. If it is run by a big pharma company, it incurs more than twice the operating costs of a plant run by a Western fine chemical company and seven to eight times as much as a plant in India.

When comparing the cost advantage of Asian companies, one should not rely solely on abstract numbers, such as labor and operating costs. As a matter

TABLE 14.3 Plant Operating Costs

	Reactor Volume	4 m³	6.3 m³	10 m³
Big pharma	Operating	80	55	30
Western CMO	cost	40	25	10
Indian CMO	($/m³/h)	8	7	5

CMO, contract manufacturing organization.

of fact, they do not take into consideration that additional costs incur, if Western companies source fine chemicals from Asia. This is especially the case if new sources of supply are developed. These include the costs for auditing and qualifying the new supplier and the registration of the product. Thus, the "total cost of ownership" can be substantially higher than the "catalog cost" of a particular product, especially if the product is new and the purchasing volume is low (see Section 12.5 and 16.1).

Western executives sometimes speculate that rapidly increasing Asian labor costs are going to close up toward Western levels in a few years' time. A closer scrutiny shows that this is wishful thinking: Taking the numbers given in Table 14.1 for Chinese and U.S. labor costs, namely $4000 and $40,000 per year, respectively, and assuming a 10% annual growth rate of the difference between them (e.g., 13% per year for China, 3% per year for the United States), it will take 25 years to reach equivalence. Even at 20% per year wage inflation, more than twice as fast as during the 2005–2009 period, the alignment will still take a dozen years! As long as there are hundreds of millions of Chinese and Indians subsisting on a monthly disposable income of less than $10, high pay increases are unlikely to happen any time soon. Furthermore, one should remember that the Asian plants were built at a time when labor costs (including construction costs) were still low—with a positive effect on the "CE" denominator in the term "RoCE" (return on capital employed).

Asia's importance is most evident in active pharmaceutical ingredient (API)-for-generics (see Section 17.1). More than 50% of the global supply originates from Asia, either in the form of bulk material or increasingly as formulated generics. The rapidly escalating share of drug master files (DMFs) submitted by Asian companies, India: 1485 (of which 78% after 2000); United States: 981 (of which 45% after 2000); Italy: 627 (43%); China: 475 (69%) clearly confirm the increase of their market share. Filings from Italy, once the prime producer, decreased sharply from 2002 onward.

> DMFs are generic dossiers filed with the FDA in order to allow the API to appear in marketed drugs. Thus, an API manufacturer files just one application for a product that can then be used to support approval of any generic based on that API.

Western life science companies so far have been reluctant to transfer sensitive technology to Asia, allowing their European suppliers to maintain an important position in the attractive custom manufacturing business for regulated intermediates and active substances for originator drugs and agrochemicals. However, as they gain confidence in the reliability and trustworthiness of the top Indian companies, direct business relationships, bypassing the Europeans, are becoming common. Many big life science companies create "Indian hubs" to manage local procurement. A selection of numbers that show the growth of the Chinese and Indian fine chemical API production is given in Table 14.4.

TABLE 14.4   Global Merchant API Production, Market Share, and Growth

Country	API for Patented Drugs			API-for-Generics		
	Sales ($ billion)	Market Share (%)	CAGR 2004– 2008 (%)	Sales ($ billion)	Market Share (%)	CAGR 2004– 2008 (%)
Europe						
Italy	0.91	6	6	2.84	17	4.3
Spain	0.28	1	6	1.05	6	4.2
Rest	8.01	42		1.71	10	3.2
Total EU	9.20	48	3	5.60	33	3.9
North America						
United States	4.25	22%	4.7	—	—	N/A
Rest	0.45	2%		—	—	N/A
Total N-A	4.70	24%	4.7	—	—	N/A
Asia-Pacific						
Japan	3.30	9	2	—	—	N/A
China	0.51	3	33	6.43	38	16
India	0.38	2	21	2.27	13	11
Rest	0.31	10		0.61	4	
Total A-P	4.50	24	6	9.31	55	14
Rest of the World (ROW)						
Total ROW	0.60	3	5	2.09	12	6
Grand Total	19.0	100	4	17.0	100%	9

Notes: (1) Sales and market share numbers refer to the year 2008.
(2) There is a noticeable disparity between the total size of the merchant API market shown in this table, namely, $36 billion, and the one shown in Tables 9.2 and 9.3, namely, $23 billion. The $36 billion number is based on the analysis of the profiles of the most significant API manufacturers made by CPA. It might contain some double counting and mixing up of drug substance and drug product (i.e., formulated APIs).
CAGR, compound annual growth rate.

Source: Chemical Pharmaceutical Generic Association (CPA), *The World API Market*, Milan 2009.

As shown in the table, the Asian fine chemical industry has a 24% share of the global market for APIs for patented drugs and 55% of API-for-generics. In sharp contrast, the United States, which constitutes the most important market for drug products, only produces 22% of the global supply. Europe, with its long-standing tradition in fine chemical custom manufacturing, holds almost 50% of the global API for patented drugs market. On the other hand, the market shares of the once-dominating API-for-generics producing countries, Italy and Spain, have fallen to 17% and 6%, respectively.

All in all, a new world order for the supply chain of pharmaceutical fine chemicals (PFCs) is emerging. In terms of geographic distribution, the center of activity will shift from the West to the East, namely from the United States to Europe and further to China and India. In terms of company structure, midsize pure players run by the founder shareholders will replace fine chemical divisions of large, publicly owned conglomerates. The global supply chain for PFCs will develop as follows.

At *present*, global life science companies mainly produce the active ingredients of their drugs and agrochemicals captively. They source advanced, exclusive intermediates from European fine chemical companies. Western fine chemical companies source basic, nonregulated intermediates from China and India. A description of the present status of PFC sourcing from Asia by Western "big pharma" companies is the following statement from a Pfizer spokesperson: "Pfizer is working intensively with Asian-based companies to move there the manufacturing of old APIs (commodities) because of cost-containment goals, while the European-based companies are mostly used for contract manufacturing of advanced intermediates for New Chemical Entities."

An example that illustrates the situation with Western fine chemical companies is Saltigo. The company reported that its share of raw materials sourced from Asia has increased dramatically since 2004, when nearly 90% of purchases were made in Europe, but this proportion fell to about 50% in 2008. While the company does not use suppliers from the region for sensitive projects, it sources catalog materials through local organizations that it has created in Shanghai and Mumbai. Cost savings have been measured at greater than 30%. The one downside, according to a key official, is that one must carefully oversee its Asian suppliers, which carries additional costs (see also Section 16.1).

In the *future*, global life science companies will reduce in-house production of APIs for patented products in favor of more outsourcing, mainly from India, bypassing their traditional European suppliers. Indicators of the *things to come* are half a dozen contract manufacturing agreements that Indian fine chemical companies have concluded with Western big pharma companies (see Table 14.5). Active ingredients for generics and even formulated products are increasingly sourced from Chindia. Indian fine chemical companies source basic, nonregulated PFCs from China. Or, more bluntly: "China is #1 in drug substances, India is #1 in drug products."

In the not too distant future, the domestic pharmaceutical and agrochemical markets in China and India will become comparable in size to those in major

TABLE 14.5 Contract Manufacturing Agreements between Indian and European Fine Chemical Companies

Indian Partner	European Partner	Manufacturing Agreement
Cadila Healthcare	Altana Pharma[a]	Two Protonix (pantoprazole) intermediates
Dishman Pharma	Solvay Pharma[b]	Teveten (eprosartan maleate) intermediate
	AstraZeneca	Nexium (esomeprazole) intermediate
	Merck & Co.	Cozaar (losartan) intermediate
Shasun Chemicals	GSK	Zantac (ranitidine) API
	Eli Lilly	Nizatidine, metohexital, and cycloserine APIs

[a] Acquired by Nycomed in September 2006.
[b] Acquired by Abbott in February 2010.
Source: Citigroup Analyst Report, October 10, 2004.

Western countries. China, given its $8 trillion GDP and its population of 1.3 billion, consistent investment in health care, and the rise in its more affluent middle-class in particular, is expected to become the third largest pharmaceutical market by 2012, up from eighth in 2006.

14.1 WESTERN HEMISPHERE

The *United States's* exodus from the global fine chemicals arena is almost complete. Although the United States is the largest user of fine chemicals in the world, such former key players in the fine chemical sector as Dowpharma, Eastman Chemical, Great Lakes, Honeywell, PPG Fine Chemicals, Reilly Industries, and Solutia all have opted out. The remaining companies, notably Ampac Fine Chemicals (formerly Aerojet Fine Chemicals), Cambrex, and Sigma Aldrich Fine Chemicals (SAFC) are the exception rather than the rule. The winners differ from the losers above all by the much bigger significance of fine chemicals within their activities. The reasons why the United States as a whole is not well positioned for fine chemical manufacture are discussed below.

European companies, too, had their share of misfortunes with U.S. subsidiaries. About a dozen fine chemical plants, which had either been purchased from domestic companies or—in a few cases—built from scratch, were later on divested or shut down (see Table 14.6).

The failure of these transatlantic expansions can hardly be just coincidental. They must rather be the result of a number of serious misjudgments:

TABLE 14.6 Divestitures of U.S. Fine Chemical Plants by European Chemical Companies

European Parent Company		Divested U.S. Fine Chemical Plant	
Company	Location	Former Owner	Location
Acino (*ex* Schweizerhall)	Switzerland	[a]	Greenville, SC
Bayer	Germany	ChemDesign	Marinetta, WI
Borregaard	Norway	polyOrganix	Newburyport, MA
DSM	The Netherlands	Catalytica	Greenville, NC
		Wyckoff	South Haven, MI
Evonik-Degussa	Germany	Proligo	Boulder, CO
Lonza	Switzerland	[a]	Bayport, TX
		Cyclo Products	Los Angeles, CA
		GlaxoSmithKline	Conshohoken, PE
Rhodia Pharma Sol.	France	DuPont	Deepwater, NJ
Siegfried	Switzerland	N/A	Carlstadt, NJ
Zach Systems	Italy	PPG	LaPorte, TX

[a] Grassroots plant.

1. It was assumed that that producing close to large U.S. pharma companies rather than in distant Europe would be a valuable competitive advantage.

2. It was not realized that the United States, with its "big is beautiful" culture, heavy regulatory burden, lack of skilled workers, dichotomy between research chemists and chemical engineers, and high cost structure was *not* a suitable region for fine chemical production.

3. Management mistakes included delegating underqualified managers, assigning profitable projects primarily to the company's homeland plants, "overinvesting" and failing to implement a "lean production" culture. Forthcoming new regulations that will impose extra costs on U.S. companies include plant security provisions to avoid terrorist attacks and Sarbanes-Oxley legislation in finance.

Europe, the cradle of the fine chemicals industry, is struggling to maintain its leading position against the massive inroads of the "high-skill/low-cost/high-future" Asian companies in a market with modest growth. High labor costs, inadequate and heterogeneous structures, insufficient number of new university graduates in chemistry and chemical engineering, and stringent safety, health, and environment (SHE) regulations all negatively impair the competitiveness.

The implementation of the REACH (*R*egistration, *E*valuation, *A*uthorization and restriction of *Ch*emicals) regulations (see also Section 5.2) will further increase manufacturing costs. About 30,000 existing chemicals will have to be registered EU-wide within 11 years at a direct cost between $3.5 and $6.5 billion. It is estimated that the larger chemical companies will be obliged to register about 1000 chemicals each. Sixty percent of these belong to the low- to mid-volume range, where the impact of the cost would be particularly felt. For fine chemicals produced or traded in the range of 10–100 tons/year, about 30 tests (physicochemical properties, toxicity, mutagenicity, and ecotoxicology) will have to be carried out. The total cost amounts to about $350,000 per product. Within the European fine chemical industry, the CROs and laboratory chemical suppliers (see Sections 2.2 and 2.3) will be particularly affected, as the registration costs have to be borne by a small volume of products, and the number of laboratory chemicals could be severely reduced.

> With the downturn in business, globalization, and political haggling over REACH, Europe did not do terribly well.
>
> —Guy Villax, Chief Executive of Hovione

> Europe's fine chemical industry needs to reinvent itself—or slip quietly into obscurity and eventual oblivion.
>
> —Rob Bryant, Brychem

Another concern relates to the import of APIs from China and India, which are estimated to account for 60–70% of all active ingredients used in generics. They are sourced from thousands of plants. The vast majority of these have not been inspected with regard to compliance with Good Manufacturing Practice (GMP) rules. Not having an elaborate QA/QC system in place gives them an unfair competitive advantage. This situation should have improved with an EU directive that came into effect on October 30, 2005. After this date, all APIs going into drugs sold in Europe must be from GMP-compliant sources. This will be comparable to the situation in the United States, where the API producer has the responsibility for GMP compliance and the FDA does the checking via inspections, of which there are about 10,000 per year. European fine chemical companies are still concerned, however, that the new directive lacks teeth and represents little movement toward consistent and uniform enforcement, especially as the EU has no foreign inspection service. Last but not least, the strengthening of the euro versus the dollar (which is also the reference currency for China and India) puts European companies at an additional disadvantage.

The present strategy of many European fine chemical companies is to concentrate domestic production on advanced intermediates and active substances for on-patent pharmaceuticals and agrochemicals. Nonregulated agro and PFC intermediates are sourced from India and China, either by straight purchasing, manufacturing agreements with selected partners, or from joint ventures (JVs), following the rule "if you can't beat them, join them." The *modus operandi* should allow the European companies to reduce total cost and thus maintain their customer base. The collateral effects are a valuable gain of know-how for the Asian business partners on the one hand, and a further reduction of capacity utilization and thus an increase in the unabsorbed fixed costs for the products, which still are manufactured in-house.

14.2 EASTERN HEMISPHERE

Chindia's involvement in fine chemicals goes back to modest beginnings. After World War II, the local chemical industry initially produced only nonregulated intermediates and API-for-generics for the local life science industry. Alongside with the expansion of global trade, both the volume and sophistication of the exported fine chemicals increased.

Nowadays, there are thousands of small fine chemical companies in Chindia. They have sales in the range of a few million to a few 10 million dollars per year. By contrast, the sales of the leading players, after having enjoyed double-digit growth rates for many years, have generally crossed the $100 million per year line (see Fig. 10.1). They are now midsize companies, which have a competitive advantage over fine chemical units of behemoths, regardless of their location.

> An example in case is Nicholas Piramal. Whereas its Pharma Solutions (CM) business had sales of about $35 million in 2003, it reached $160 million in the fiscal year 2009/2010.

Chindia's export volumes of regulated intermediates are still modest. China, in particular, is mainly exporting mature low-price/large-volume type PFCs. Western fine chemical and ethical pharmaceutical companies typically source not more than 10% of their chemical requirements from there.

> An example is Syngenta, the world's number one agrochemical company. While Syngenta's purchasing volume from Asian Pacific countries almost quadrupled in the 2001–2006 period, it is still relatively small, namely 16%, as a percentage of total "raw materials for synthesis" (China 10%, India 3%, Japan 3%).
>
> *Source*: Syngenta 2006 Supplier's Day.

For exclusive products, supplied under secrecy agreement, observance of intellectual property rights (IPR) still is a concern of the Western business community. China joined the World Trade Organization (WTO) in 2001 and has since amended its laws to comply with the WTO agreement on Trade-Related Aspects of Intellectual Property Rights (TRIPS). This now includes also chemical and pharmaceutical products. India, following the entry into the IPR regime from January 1, 2005, is moving forward to include advanced intermediates. However, although pertinent legislation exists, enforcement is not always up to the expectations of the Western business partners. In order to understand the attitude of the Asian business partners, one must bear in mind that intellectual property (IP), unlike the ownership of land and real estate, is not part of their cultural heritage. Thus, IP rights are accepted rather reluctantly and on an opportunistic basis.

CIPLA (Chemical Industrial and Pharmaceutical Laboratories) is the world's largest supplier of anti-AIDS drugs. With Triomune, launched in 2000, it reproduced the most important cocktail of antiretroviral drugs, Stavudine, Lamivudine, and Nevirapine, sold it for $1 a day to "Médecins sans Frontières" ... and engaged in a fierce battle with the patent holders. Richard Sykes, head of GlaxoSmithKline, denounced Yusuf Hamied, president of CIPLA, as "pirate," Hamied fired back saying, that GSK was a "global serial killer" for charging such high prices

Another example in case is the anticancer drug Gleevec (imatinib) from Novartis. Although the molecule was known for a long time, Novartis obtained patent protection in Western countries for a specific crystalline modification. Indian courts rejected the application on the grounds of an insufficient level of innovation.

Current patent legislation still allows China and India to develop APIs for existing proprietary drugs and sell them on their rapidly growing home markets. After patent expiration, they can immediately serve the Western market places. Whereas they initially focused on producing simple, nonregulated fine chemicals and API-for-generics for countries that did not check regulatory compliance (this included Europe until recently), they now enter into advanced intermediates that have to be produced under cGMP regulations.

There is much debate about the question of which of the two emerging superpowers, China or India, is preferable as business partner for Western life science and fine chemical companies. India has a democratic constitution, is more advanced in technology, has been involved in PFCs longer, and has a legal system based on English law. It has left behind its heritage of bureaucracy, corruption, and protectionism to a large extent. Also, with respect to adopting modern business models, Indian fine chemical companies make rapid progress. Figure 14.1 illustrates the offering of Hikal Ltd., a leading Indian fine chemical company. It gives a good insight in the company's business model, which is fully up to the one of its Western competitors.

China, meanwhile, has an even lower cost structure than India, is more advanced with regard to infrastructure (ground and air transportation, electric power, telecommunications, etc.), and has well-developed technology parks. China is educating engineers, whereas India is better in generating excellent process development chemists. On the downside, it has a different legal system, a serious language barrier, and a legacy of regulatory, environmental, health, and safety problems. Also, there is a government-imposed mandatory technology transfer for foreign companies building plants in India.

In order to attract direct foreign investments, both countries have designated about 20 Special Economic Zones (SEZ), geographical regions that have more liberal economic laws and incentives than the rest of the country. The most successful SEZ in China, Shenzhen, has developed from a small

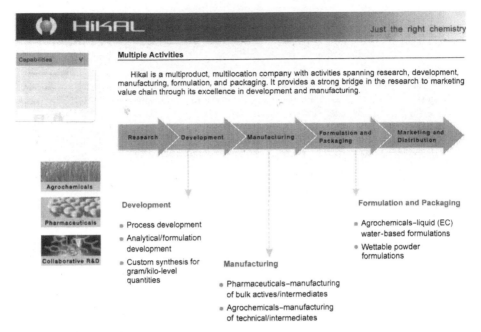

Figure 14.1 Offering of Hikal Ltd. EC, emulsifiable concentrate.

village into a city with a population over 10 million within 20 years. In India, the SEZ in Aurangabad, Maharashtra, provides infrastructure for the pharma and biotech industries. China's flagship high-tech park Zhangjiang in Pudong, Shanghai hosts R&D facilities of 320 life science companies, of which 29 are global pharma, with 20,000 scientists.

In order to get a stronger grip on the Western markets, cash-rich Indian pharmaceutical and fine chemical companies entered into an acquisition spree of Western fine chemical and generics companies. Between 2004 and mid-2006, more than 20 deals were completed (see Appendix A.10 and Section 2.5). Like their European counterparts, which had a negative experience with their transatlantic expansions (see Table 14.6), not all Indian overseas acquisitions were a sweeping success. Nicholas Piramal acquired two plants in the United Kingdom, one in Morpeth, from Pfizer, and another in Huddersfield from Avecia. The workforce of 450 at Morpeth had been reduced by 15% in 2008, and the site at Huddersfield was closed in 2009. Likewise, Shasun had acquired two UK API plant in Dudley and Annan from Rhodia Pharma Solutions. The Annan plant was closed because of "underutilization" and was later sold to Phoenix Chemical. Conversely, Western fine chemical companies expand to Asia (see Appendix A.11).

14.3 COOPERATION MODELS

Apart from conventional supply contracts (see Section 12.7) there are a number of models for a closer global distribution of tasks between the life science and fine chemical industries. Their common intent is to find an equitable risk/reward balance between the global partners. The models differ mainly with regard to the degree of financial participation and corporate integration. The top-tier fine chemical companies have already adopted an "integrated-service offering" approach providing a complete suite of services related to the life science molecule from early process development to commercial manufacture and moving products through development to commercialization quickly and cost-effectively. The concepts, their pros and cons, and practical examples are described below.

In an *umbrella manufacturing agreement*, a framework is set up covering supply contracts for specific fine chemicals to be manufactured exclusively. The underlying concept of is that, along with the globalization of the fine chemical industry, Asia will take the lead in intermediates and actives for off-patent drugs and agrochemicals, as well as early stages of proprietary drugs and agrochemicals, while Europe and the United States will concentrate on the advanced steps, which require full cGMP compliance and contain sensitive IP. The benefits for the customers of having their Western suppliers offshore early stages of fine chemical manufacture are cost reductions without compromising the quality, trustworthiness, and IP. The model is attractive for Asian companies as long as they lack direct access to the global life science industry, and as long they can benefit from the Western know-how in technology transfer. In both areas, Chinese and Indian fine chemical companies are making rapid progress. The challenge is formulating an equitable profit sharing. Umbrella manufacturing agreements are the typical legal form for executing a preferred supplier agreement. Many agreements of this kind exist between big pharma and leading fine chemical companies.

The *build, operate, and transfer* model is borrowed from the automotive industry, where parts suppliers install their plants as satellites on the premises of the final assembler of the car. The main advantages are a "zero distance" supply chain and shared infrastructure. Thus, the supplier builds, and later on operates, a multipurpose plant dedicated exclusively to manufacturing the customer's products. In addition, he provides staff, QA/QC, maintenance, and other support. The customer gets the control of the production and has the option to purchase the plant after a certain period of time. As the service fee is not dependent on capacity utilization, the risk for the supplier is limited, but the profit will also be lower as compared with a fully utilized autonomous plant.

Genentech of the United States, which is now part of Roche, has excised an exclusive option to acquire an 80,000-L mammalian biopharmaceutical production facility in Singapore from Lonza, Switzerland in 2009. The price for the transaction was $290 million plus $70 million in milestone payments. Lonza signed a similar agreement with Novartis for the process development, scale-up, and industrial-scale production of APIs for biopharmaceuticals in mid-2008. This "pipeline contract" variety foresees initially the erection of the (empty) production building plus infrastructure at Lonza's expense alone (equaling about 15% of the total investment [which can cost up to $500 million]) and the construction of the production lines "on demand." The advantage of this solution is the reduced response time (insertion of production lines saves about three out of the five years needed for the building of a grassroots plant) and closely aligns the investment expenditure to sales proceeds.

Syngene, an Indian CRO, has built a dedicated R&D laboratory, which employs 400 scientists. It is operated on behalf of Bristol-Myers Squibb (BMS) until at least 2013, when BMS has the option to purchase it. BMS says that if it had decided to build its own Indian lab lacking the experience of undertaking a large project in India, it could not have taken full advantage of the low costs.

A variety of the build, operate, and-transfer model is the *sale of an existing fine chemical production unit*. It allows the customer to take immediate advantage of production capacity.

Siegfried of Switzerland sold a drug formulation unit for CHF 47 million (approximately $43 million) within its main Zofingen site to Arena Pharmaceuticals of San Diego in the first half of 2008. Simultaneously, a conventional supply agreement for the custom manufacturing of the antiobesity drug Lorcaserin Hydrochloride (Phase III) was concluded. This is the second such deal by Siegfried after having sold a PFC manufacturing unit for CHF 55.5 million (approximately $51 million) to Celgene of the United States in December 2006. The facility, also part of the Zofingen site, has the capability of producing multiple drug substances and is used to produce Revlimid (lenalidomide) and Thalomid (thalidomide), as well as APIs for drugs under development. Siegfried has the right to use the unit for its own product needs when it is not needed by Celgene.

As a further, modern alternative, a fine chemical company assumes the operation of a pharma plant, whereby

- the ownership remains with the pharma company.
- the fine chemical company assumes the life-cycle management of the API(s) produced.
- it lets the pharma company take advantage of the learning curve, thus allowing it to capture most of the generic market.

It is granted the right to sell the API(s) to third parties after patent expiration.

A *JV* allows a combination of tangible and intangible assets of the partners in a way to optimize the combined offering. JVs are popular in the heavy chemical industry, particularly between Western and Chinese companies in

China. However, the failure rate of JVs is quite high. This is particularly the case for horizontal JVs, where all activities, such as R&D, production, and M&S, are shared. The main reasons are culture clashes that prevent the expected synergies and conflicts of interests with other activities of the partners. Actually, the joint venture status of both examples shown hereafter was superseded by sole ownership after a few years.

> Evonik-Degussa and Lynchem established a joint venture for exclusive synthesis in Dalian, China, by acquiring 51% of the shares from its Chinese partners, Yuncai Wang and Jingkun in 2006. The rest of the shares were purchased in 2008. "In taking over the remaining shares, we are strengthening our global exclusive synthesis business and further expanding it in line with our successful concept of horizontal integration," Evonik executive board member Alfred Oberholz said at the time. Now, the site has been closed because of environmental problems.

> Hovione of Portugal decided to enter the market for generic APIs for X-ray contrast media in early 1993. Modifying conventional manufacturing processes had enabled the first steps to be done at contractors in China with Hovione technology as early as 1994, and the critical GMP steps took place in its own plants. In order to win the up-hill fight against the established competition from the innovators (notably Bracco of Italy, Nycomed (now GE Health), and Nihon-Schering in Japan), a JV was concluded with Hysin of China around 2000. The JV allowed Hovione to penetrate the Asian market, limiting risk and avoiding costly capacity expansions, such as for handling iodine. Full ownership of the 300-m^3 Hysin plant was taken in 2008.

A *financial participation* allows the acquiring party to become a shareholder in its partner. Depending on the percentage of the equity purchased, a minority or majority interest in, or even a full ownership of, the company is obtained. The high market capitalization of Indian fine chemical companies, with earnings before interests, taxes, amortization, and depreciation (EBITDA) multiples of 15 and price-equity ratios (P/Es) typically above 20, basically excludes this model for potential Western partners. It is, however, of great interest to Asian companies eager to get a foothold in the West, allowing them to assess the missing link in their offering, namely advanced PFCs for new drugs.

CHAPTER 15

Biotechnology

As described in Section 4.2, traditional biotech processes, namely biocatalysis and microbial fermentation, are used for the production of small molecules, whereas the modern cell culture methodology allows the production of high-molecular-weight (HMW) biopharmaceuticals. A growth rate of 10–15% per year is expected for the biotechnological segment, while the average increase of the pharmaceutical market is expected to soften to less than 5% per year. In terms of technologies, the demanding mammalian cell cultures are expected to grow fastest, followed by microbial fermentation.

15.1 SMALL MOLECULES (WHITE BIOTECHNOLOGY)

The perspectives for an increasing use of white biotechnology techniques, that is, biocatalysis and microbial fermentation, for low-molecular-weight (LMW) chemicals are promising. The substitution of traditional chemicals by biotechnological processes constitutes the most important means for reduction of manufacturing cost for fine chemicals. As white biotechnology is very eco-friendly, its more widespread use will also have a positive effect on the carbon footprint of the chemical industry. Already more than 30% of fine chemical production processes comprise at least one biotechnology step. It is expected that revenues from white biotechnology will account for 9%, or $225 billion, of the global chemical market of $2,500 billion by 2012. Biofuels, in particular bio-ethanol and bio-diesel, will generate 40% of the anticipated sales, followed by plant-derived materials, for example, cellulose, active pharmaceutical ingredients (APIs) ($25–30 billion), chemicals and polymers, food and feed and enzymes.[1]

---

[1] *Sources*: Stanford Research Institute, F.O. Licht, Frost & Sullivan, McKinsey.

The perspectives for the future are promising: In 10–15 years, it is expected that most amino acids and vitamins and many specialty chemicals will be produced by means of biotechnology.

—BASF news release

15.2 BIG MOLECULES (BIOPHARMACEUTICALS)

Cell culture, particularly mammalian cell culture, processes enable the production of new, hitherto unknown drugs, the biopharmaceuticals, to take place. Whereas the first biopharmaceuticals appeared in the 1980s, the year 2001 marked a turning point, insofar as it was the first time that more new biological entities (NBEs) than new chemical entities (NCEs) were approved by the FDA. Today, biopharmaceuticals account for about $55–$80 billion, or 10–15% of the total sales of the pharmaceutical industry. Although their share will increase further, they are unlikely to ever fully replace their traditional counterparts. In many applications, small molecules will remain the drugs of choice. Biopharmaceuticals are mostly made by mammalian cell culture technology. Its main disadvantages are low volume productivity and the animal provenance. It is conceivable that other technologies, particularly plant cell production, will gain importance in the future.

A brilliant example of the industrial-scale application of plant cell fermentation is the new process for the production of the anticancer drug paclitaxel developed by Bristol-Myers Squibb (BMS) (see Fig. 15.1). It starts with clusters of paclitaxel producing cells from the needles of the Chinese yew, *Tsuga chinensis*, and was introduced in 2002. The API is isolated from the fermentation broth and purified by chromatography and crystallization. The new process replaces a previously used semisynthetic route. It started with 10-deacetylbaccatin (III), a compound that contains most of the structural

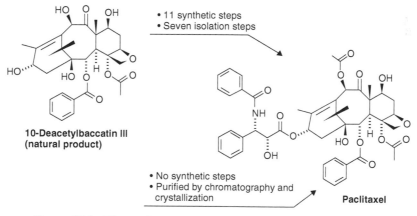

Figure 15.1 Plant cell fermentation process for Taxol (paclitaxel).

complexity of paclitaxel and that can be extracted from leaves and twigs of the European yew, *Taxus baccata*. The chemical process to convert 10-deacetylbaccatin (III) to paclitaxel is complex. It includes 11 synthetic steps and has a modest yield. A plant cell process is also in the making for insulin, demand for which is expected to reach 12,000 kg by 2012. The Canadian firm SemioSys Genetics, which is developing the process based on safflower, anticipates capital costs of 70% and product costs of 40% as compared with exiting insulin production relying on genetically engineered yeast (*Saccaromyces cerevsiae*) or *Escherichia coli*.

The pros and cons of an involvement of a fine chemical company in cell culture technology are listed below:

Pros:

- Large-molecule drugs should see a major boost in relative importance over small-molecules drugs within the next 5 years. Whereas just 1 out of the world's top 10 drugs was a biopharmaceutical in 2001, the number went up to three in 2009 (see Table 11.3), and is expected to increase further to eight by 2016 (see Table 15.1). Abbott's anti-inflammatory Humira (adalimumab) is only the third drug after Lipitor and Plavix expected to reach sales in excess of $10 billion. All in all, the sales of biopharmaceuticals are growing by 15% per year, that is, three times faster than LMW drugs, and are expected to pass the $150 billion per year threshold by 2015. The custom manufacturing market is expected to reach $5 billion by 2015.

- The likelihood of developing a new biopharmaceutical successfully is significantly greater than in traditional drug development. Because interactions, side effects, and carcinogenic effects are rare, 25% of biopharmaceuticals that enter Phase I of the regulatory process eventually are granted approval. The corresponding figure for conventional drugs is less than 6%.

- The traditionally large share of outsourcing.

- Small number of custom manufacturers with industrial-scale manufacturing capabilities in this demanding technology, primarily Boehringer-Ingelheim of Germany, Lonza of Switzerland, and Nicholas Piramal of India (through the acquisition of a former Avecia operation) as opposed to several thousand fine chemical companies using traditional chemical synthesis.

- Asian competition is lagging behind. Examples in emerging companies are the joint venture (JV) between AutekBio and Beijing E-Town Harvest International in China (which was described as the largest biopharmaceutical contract manufacturing organization in Asia) and between Biocon in India and Celltrion in South Korea.

- Same customer category: life science, especially the pharmaceutical industry

TABLE 15.1   Top 10 Drugs 2016

Proprietary Name		Generic Name	Company
High Molecular Weight (Biopharmaceuticals)			
1	Humira	Adalimumab	Abbott
2	Enbrel	Etanecerpt	Amgen
3	Prolia	Denosumab	Amgen
4	Rituxan	Rituximab	Roche/Biogen Idec
5	Avastatin	Bevacizumab	Roche
6	Herceptin	Trastuzumab	Roche
7	Remicade	Infliximab	J&J/Merck & Co.
8	Lantus	Insulin glargine	Sanofi-Aventis
Low Molecular Weight (Conventional Chemical)			
1	Crestor	Rosuvastatin	AstraZeneca
2	Advair/Seretide	Salmeterol/fluticasone	GSK

Source: EvaluatePharma.

- Similar business types: custom manufacturing of proprietary drugs, biogenerics
- Similar regulatory environment: FDA regulations, especially Good Manufacturing Practice (GMP)
- Basically similar manufacturing processes: raw material preparation, reaction, isolation, purification, workup, etc.
- Existing infrastructure (utilities, maintenance shops, warehouses, etc.) can be used
- Growing demand, lack of suitable production capacity, limited competition, and thus, a seller's market

Cons:

- High entry barriers because of demanding technology. The construction of a large-scale plant for the production of biopharmaceuticals by cell culture fermentation costs around $500 billion, takes 4–6 years, because it is so technologically, legally, and scientifically demanding. Therefore, it must be planned even before Phase II studies begin.
- In terms of process technology, biotechnology differs substantially from traditional chemical synthesis. Biopharmaceuticals cannot be produced in conventional multipurpose fine chemical plants. As shown in Table 4.3, the specifications of the two plant and process types are almost totally different. This raises the entry hurdle even higher.
- High financial exposure (see Table 4.3): (1) high capital intensity ("massive investments are needed at a time when chances of success are still very low" [see reference [5] in part I]) and (2) risk of batch failures (contamination).

- Unlike the biopharmaceutical start-ups, the emerging big biopharmaceutical companies are adopting the same opportunistic outsourcing policy as big pharma companies. Thus, Amgen, Biogen/Idec, Eli Lilly, Johnson & Johnson (J&J), Medimmune, Novartis, Roche/Genentech, and Wyeth have begun investing heavily in in-house manufacturing capacity. With three plants in the United States, two in Japan, and one each in Germany and Switzerland, Roche has the largest production capacity. The trend is also favored by the fact that cell cultures produce the API directly. Intermediates are not isolated. This goes against the industry's preference for keeping the last synthetic step in-house. Overall, the ratio between captive and third-party manufacturing is approaching the 60/40 ratio prevailing in the traditional pharma industry.
- New developments in expression systems for mammalian and plant cell technology could reduce capacity requirements substantially. Actually, the titer in large-scale mammalian production, which doubled from 1–2 g/L to 2–3 g/L between 2005 and 2010, is expected to double again to 5–7 g/L by 2015 and once more to 10 g/L by 2020. Furthermore, the widespread application of "single-use disposable bioprocessing technology" is considered by experts as "the hottest buzz in town." It advantageously substitutes for stainless steel production trains, at least for short production campaigns.
- New transgenic production systems are emerging that possess the potential to become industrially successful (e.g., transgenic moss, lemna, fungal, or yeast expression systems, transgenic animals and plants, such as tobacco plants).
- Legislation and regulation of biotechnology is not well defined yet and leads to differences in interpretation and other uncertainties. In the United States, legislation is not yet in place for biosimilars, the generic counterpart of generics in small-molecule pharmaceuticals.

The inherent risks of the mammalian cell technology lead several companies to opt out of mammalian cell technology or to substantially reduce their stake. Examples are Cambrex and Dowpharma in the United States; Avecia, DSM, and Siegfried in Europe; and WuXi Pharma Tech in China.

In conclusion, biocatalysis should be, or become, part of the technology toolbox of any fine chemical company. Cell culture fermentation, on the other hand, should be considered only by large fine chemical companies with a robust financial backup and a long-term strategic orientation.

CHAPTER 16

Ethical Pharmaceutical Industry

Within the next 5 years, the pharmaceutical industry will be facing the severest challenges since Bayer started marketing acetyl salicylic acid worldwide under the brand name Aspirin in 1899. The three major negative trends are the "patent cliff," the low innovation rate, and government pressure on drug prices. The patent cliff results from an exceptionally high number, about twice as much as currently, of new products, which had been launched in the period 1997–2002, losing patent protection. Between 2015, drugs with aggregated sales of $140 billion (in 2008) will fall over the "cliff," causing the total market to shrink by about 16%. Several companies will lose protection for their most lucrative products. According to IMS Health, Pfizer/Wyeth will be the worst hit. Because of the forthcoming patent expiration of Lipitor, revenue losses of 3% in 2011, 28% in 2012, and 35% in 2013 are anticipated. The impact on Novartis will be 2% in 2011, 13% in 2012, and 35% in 2013, when the Diovan patent will expire. Other big pharma companies affected include Eli Lilly (Zyprexa), Bristol-Myers Squibb (BMS) and Sanofi-Aventis (Plavix), and Astra Zeneca (Seroquel).

> In some rare cases, patent-holding big pharma companies manage to retain an important market share after patent expiration, thanks to process improvements and/or transfer of the production to low-cost countries, which result in dramatically lower API costs. An example in case is Pfizer's large volume (>1000 tonnes per year) analgesic Neurontin (gabapentin). Shortly after patent expiry, the market share dropped to 15%, but could be augmented to 75% after the API had been outsourced to an Indian producer, which had developed an innovative, cost-effective process.

Now that the "low-hanging fruit" (cures for acute diseases) have been harvested, the innovation rate is only about half of what it used to be during the so-called "golden years." Pharma is unable to compensate this loss with new drug launches and a negative trade balance of about $10 billion per year is likely to result.

TABLE 16.1   Government Provisions for Curbing Health-Care Costs

1	"Generic switch"	Obligation of physicians to prescribe generic equivalents to proprietary drugs after patent expiration
2	Price control	Price control on prescription drugs. Japan, the world's second largest pharma market, has led the way by imposing yearly drug retail price reductions
3	Parallel imports	Parallel imports (from countries with lower drug prices) are permitted
4	"Pay-for-performance" ("fourth hurdle")	Authorities in many countries no longer accept high prices for new drugs with only marginal therapeutic benefit improvements. The British "NICE" already has included economic considerations in their new drug approval process; health authorities of other countries might follow suit in adapting "pay-for-performance" criteria

The underlying figures are that the industry will have to absorb about 75%, that is, $20 billion, of the revenue losses of $28 billion per year due to patent expirations and that 20 new products launched every year will generate sales of 20 × $500 million, equaling $10 billion. It all results in a negative industry growth of $10 billion per year.

On top of these two internal problems, the industry is also faced with increasing pressure from governmental authorities. As major part of efforts to curb escalating health-care costs, they want the industry to reduce drug prices. Table 16.1 lists four measures eyed by the authorities in many countries. The health-care reform introduced in the United States in early 2010 will cost drug companies $85 billion in lower prices they will receive from government programs and surcharges. Conversely, drugmakers ultimately stand to gain when more than 30 million previously uninsured people begin using prescription medicines. Brand-name drug firms have won some advantages over rival generic drug producers. They also avoided a new provision that would have banned them from paying their generic counterparts to abandon patent challenges that could lead to early market entry for competing generic medicines. Biotech firms were granted a 12-year sales exclusivity for innovator biopharmaceuticals. China is considering reviewing its current policy of reimbursing many innovative drugs at western levels.

The problems of the pharma industry have not gone unnoticed by the general public. They have provoked a paradigm change in public perception—from "most admired" in the 1990s to the bottom rank of public esteem.

Big pharma increasingly face a conflict between the goals of corporate wealth and public health: it is far more influenced by Wall Street than the public's need for safe, effective and affordable medicines.

—Paul J Reider, VP for Chemistry Research, Amgen

The pharmaceutical industry is reacting to these challenges mainly by undertaking a major effort to revise strategies in operations, R&D, and marketing. The measures and their implications on the fine chemical industry will be discussed in the following section. Also, in response to its impaired image it, wants to be perceived as *health care* and not as a *chemical* industry—or, as Eli Lilly put it, from FIPCO (fully integrated pharmaceutical *company*) to FIPNET (fully integrated pharmaceutical *network*).

16.1   RESTRUCTURING AND OUTSOURCING

The total control of the whole supply chain, from discovery research and raw material sourcing all the way to postlaunch monitoring of patients, has been a key strategy for the pharmaceutical industry, particularly big pharma. This axiom is now being abolished and the industry, realizing that in-house manufacturing is negatively impacting overall return on investment and adding little, if any, competitive advantage, is switching from opportunistic to strategic outsourcing. Pertinent manufacturing strategy reviews will be part of the measures to restore profitability. Pharmaceutical companies are becoming more cost-sensitive with regard to in-house production.

In order to reduce costs throughout the value-added chain, the industry will recognize that "just make it and don't get us into trouble" is no longer a sustainable manufacturing strategy and recognize the benefits of the "buying instead of making" their drug substances.

> Other industries, such as automobile companies, have shown the way to go much earlier. They are becoming "vehicle brand owners" concentrating on the design and marketing of the cars. The car producers have reduced their share of the added value to 35%. As the parts suppliers spend more on R&D (6–9%) than the car producers (4–6%), the share of the latter will diminish further.

In pursuit of the "total control of the supply chain" axiom, the industry had been applying an opportunistic outsourcing policy with regard to chemical manufacturing of pharmaceutical fine chemicals (PFCs) until recently, despite compelling financial reasons for outsourcing brought forward by independent experts (see Section 10.2). A number of considerations supported this notion, namely quality and intellectual property (IP) concerns, abundant cash reserves, the low impact—typically less than 10%—of the cost of goods sold (COGS) in the overall cost structure of a drug, the tax advantages for the local production of PFCs offered by locations such as Ireland, Puerto Rico, and Singapore, and underutilized in-house production capacity.

Responding to the cost-cutting mantra, the industry is heavily divesting its manufacturing plants. Pfizer, notably, has a daunting synergy target of $6 billion for the period 2007–2012. The company owned 100 plants worldwide after the acquisitions of Warner-Lambert in 2000 and Pharmacia in 2003. By

2008, it had whittled that down to 43, only for the total to jump up to 79 again after the 2009 acquisition of Wyeth, which outsourced all of its small-molecule active pharmaceutical ingredient (API) manufacturing. Out of the 79 plants, 47 are manufacturing LMW while 16 are manufacturing HMW APIs. As of 2010, Pfizer is outsourcing approximately 30% of its requirements through relationships with about 250 contract manufacturers. A good portion of them, about 70%, are former Pfizer sites. AstraZeneca (AZ) intends to outsource *all* its API production over the next 5–7 years, BMS, Eli Lilly, GlaxoSmithKline (GSK), and Merck & Co. are scaling back the number of their plants by up to 50% and are planning to outsource as much as 40% of their API needs.

> [BMS] will continue to use third-part suppliers for back-up production of newer drugs and aggressively outsource manufacturing of mature products.
>
> —Lamberto Andreotti, COO of BMS

> I always dreamed-hoped that big pharma would move from purely opportunistic to strategic outsourcing of chemical manufacturing. Now, a vision has become true, albeit 20 years later than expected.
>
> —Peter Pollak, quoted by Rick Mullin
> in C&EN News, November 9, 2009, vol. 87 no. 45, p. 34

Pharma's paradigm change with regard to API sourcing is in principle positive for the fine chemical industry, both for its custom manufacturing and even more so for its API-for-generics businesses. In custom manufacturing, it will compensate the negative impact of pharma's drying up new product pipelines. However, one should not expect that the present overcapacity will disappear anytime soon. On the one hand, Western pharma companies sell underutilized plants (see Table 2.7), while on the other, new capacity is coming onstream in Asia at a rapid pace. Recent mega-mergers are only compounding the overcapacity problem. API-for-generics will also profit from the patent cliff, the government initiatives promoting the generic switch, and the rapidly increasing demand in "pharm-emerging" countries (Brazil, India, Indonesia, Mexico, Russia, and Turkey). Asian, particularly Chindian, firms will be the main beneficiaries. As confidence in their technological capabilities and respect for intellectual property rights (IPR) rules increases, Asian, especially Indian companies, are becoming important partners in outsourcing deals, and their share in the global outsourcing market is increasing. Big pharma executives unanimously state their intentions to source more from Asia. In order to do so, sourcing offices are being set up. Also, outsourcing does not increase at the same speed at all companies. If the "make-or-buy" decision for a specific API depends primarily on cash-out considerations, the custom manufacturer will lose out.

When evaluating local procurement versus offshoring, the consideration of the "total cost of ownership" is gaining more weight in sourcing decisions. Actually, apart from the "ex-works" price, a number of additional cost elements have to be considered. In the normal course of business, these include the sales terms and delivery conditions (including packaging, transportation, and custom duties), the costs of auditing and qualifying the Asian supplier, and for registering the product. Furthermore, abnormal situations, such as delayed or incomplete deliveries, handling quality complaints and IP violations have to be taken in due consideration too. Thus, the total cost of ownership of the purchase can be much higher than the invoiced price and make the Asian option unattractive. As some of the aforementioned expenses (which may also include the cost of running a procurement office in Asia) are fixed costs, the Asian solution is the less attractive, the smaller the size of the deal. It is also less attractive in the case of an exclusive product as compared with a standard product. In the latter case, IP issues and dependency on one supplier are not at stake.

It can all boil down to a trade-off between more than $30 million revenue loss per day if a $1 billion blockbuster drug runs of out stock because of a 10-day delay in delivery from an Asian supplier as compared with a few percentage points lower cost for the API.

An example in case is a leading German pharmaceutical company, which had chosen two Chinese fine chemical companies for the supply of a pivotal intermediate. When there were interruptions in shipments, it found out that the starting material for the intermediate originated from just one source, which local authorities had closed down after serious environmental problems had arisen.

All in all, Asian suppliers of fine chemicals to Western life science companies are most competitive for regular supply of large volumes of standard products and are least competitive for one-shot requirements for exclusive products.

Pfizer plans to increase its outsourcing, especially to lower cost destinations. Currently, 15% of our product manufacturing is externally outsourced and we plan to double this to 30% by 2010.

—David Shedlarz, vice chairman, Pfizer, Inc.

16.2 R&D PRODUCTIVITY

For the sake of its own healthy development, the pharmaceutical industry should ensure that the revenue drain from drugs coming off patent (see Table 17.1) is compensated for by new drug sales. Unfortunately, this is not the case by far. From an all-time high of 59 new molecular entities (NMEs), 51 of which were new chemical entities (NCEs), approved by the FDA in 1997, the number dropped to an all-time low of 18 NMEs (15 NCEs) in 2005. The situation looks even more alarming if one considers the truly innovative new drugs only. In the period from 1998 to 2002, 415 new drugs were approved altogether.

Of those, 133 (32%) were NMEs; the others were mere variations of old drugs and, of those 133, only 58 were priority review drugs. That averages out to no more than 12 innovative drugs per year, or 14% of the total (see Reference 3 in Part II). With the disappearance of the "blockbuster model", the industry is not only suffering from fewer FDA approvals, but also lower sales of newly launched drugs.

An example in case is Folotyn (pralatrexate) from Allos Therapeutics, (http://www.allos.com) approved in September 2009 for peripheral T-cell lymphoma/relapses of a rare type of blood cancer affecting 5,000 Americans per year

To be a genuine advance, a new medicine ... has to meet patients' needs better than anything else that's available

—Roche Annual Report 2005, p. 12

The agony of new drug launches is not caused by a reduction in R&D spending, which in the United States alone doubled from $32 billion in 2001 to $75 billion in 2007. This not only demonstrates the slump in pharma research productivity but also causes a dramatic increase in the development cost per new drug. The cost increased from $350 million (1990) to $500 million (1995), $750 million (2000), $1 billion (2005), and $1.5 billion (2010). Only large-selling drugs can allow the pharma industry to recover these enormous expenses.

In line with the increase in pharma R&D spending, the total number of drugs in development in the global pharmaceutical industry increased 2.4-fold from 3737 to 8832 between 2001 and 2009 (see Fig. 16.1). However, the ratio

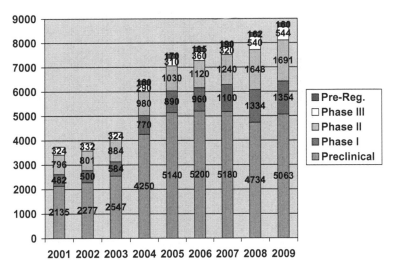

Figure 16.1   Development pharma pipeline 2001–2009.
Note: Rough estimate; biologics are excluded.
Sources: Pharma Projects, McKinsey, IMS Health.

between development drugs in Phase III compared with those in Phase I has worsened from 6.6 to 1 in 2001 to 9.3 to 1 in 2009. This 50% increase in attrition clearly confirms the slump in pharma R&D productivity. The underlying reasons are the robotizing of lead compound generation on the one hand and more demanding therapeutic targets (chronic vs. acute diseases) on the other hand. Nils Behnke of Bain Consultancy estimates that the value of drugs in development in the industry's collective pipeline that could be launched between 2010 and 2015 is just $30 billion. This is an alarming figure if compared with the $140 billion originator drugs which will fall over the patent cliff in the same period! According to other experts, cumulative drug development success rates (Phase I through approval) within major pharma have declined in recent years, from roughly 18% between 1996 and 1999 to 9% from 2000 through 2003.

The financial community is increasingly airing concerns that modern techniques for drug discovery and development, in which researches disregard natural sources and create and rapidly test a million or more chemical compounds for possible pharmaceutical activity, have failed to make good on their promise of bringing a flood of medicines to patients and profits to investors. Actually, these new technologies have yet to yield appreciable results in late-stage NMEs. There are several root causes for the slump in pharma productivity:

- The complexity of modern drug development. On its evolution from trial and error to the systematic search for new drugs, the whole process has become extremely complex. Comparing it to mechanical industry, it is like having a very well-furnished toolbox on the one hand, and a very complicated machine with an unidentified defect on the other hand. The twofold challenge is to identify the site of the defect and to find the right tool to repair it.
- Tighter, and to a large extent, self-imposed, rules for the quality of the safety profile of new drugs.
- The focus of big pharma on developing "me-too" blockbuster drugs in largely exploited therapeutic areas and neglect of acute diseases
- High entry barriers in new therapeutic categories, sometimes referred to as the "high hanging fruit," such as Alzheimer's disease, metabolic syndromes, and multiple sclerosis, which develop over many years.
- The growth in launches of NBEs has not compensated for the decline in NCEs.
- The powerful marketing engine of big pharma allows them to sell mediocre products.
- Culture shock within the R&D divisions of the mega-merged pharma companies and a brain drain of talented young researchers.
- The "double exposure" on innovation and lean production of pharma's present business model.

- New technologies for drug development have yet to yield appreciable results in late-stage compounds. Genomics are turning out great numbers of new, potentially useful drug targets; high-throughput screening makes it possible to evaluate large libraries of new molecules produced with automated laboratory reactors, but essentially, it has only created a massive amount of data so far. It will take several more years to produce commercial drugs.
- Bad side effects that typically show up in the final stages of clinical testing are related to pharmacokinetics, especially absorption, distribution, metabolism, excretion, and toxicity (ADMET). Computational models and experimental methods for a timely detection of ADMET properties are being developed. A major problem is predicting the type of liver toxicity that is not detected until *after* a drug has been marketed, as experienced in the well-known cases of Sankyo/Warner-Lambert's Rezulin/ Noscal (Troglitazone) and Bayer's Lipobay (Cerivastatin).

In the following section, six elements of modern methodology for drug development are examined:

1. *Combinatorial Chemistry (combichem) and High-Throughput Screening (HTS).* The input for *new drugs in development* is determined by the number of new molecules made available as a whole ("libraries") on the one hand and "targets" for drug action on the other. Modern combichem allows the preparation of large libraries of tens of thousands or even millions of potential lead compounds, though this still a "needle in a haystack" compared to the 10^{10}–10^{15} small molecules that theoretically can be made from the four elements C, H, N, and O. HTS allows the screening of these libraries for drug candidates. It emerged from a convergence of low-cost computer systems, reliable robotic apparatus, sophisticated molecular modeling, statistical experimental strategies, and database software tools. About a decade ago, the drug industry embraced combichem and HTS. The idea was that the rapid evaluation of large numbers of drug candidates might be easier and more efficient than the often tedious and time-consuming process of designing drugs one by one or screening compounds in small numbers and would offer a new way to discover drugs. However, both methods have turned out to be a disappointment so far. They showed few tangible results and the promised productivity enhancement has not materialized. Supporters say that the modest number of successes is due to the long development time for a new drug rather than the inefficiency of the technology. It is, therefore, too early to discredit combichem/HTS for failure to produce new drugs, they say. Their adversaries say: "You cannot beat innovation with statistics." They question whether it makes sense to discard completely the *traditional cell- and organism-based strategy* and to take the medicinal researchers out of the loop at a crucial phase in new drug discovery,

replacing their ingenuity and intellectual capacity with the programmed mechanics of robots.

2. *Genomics and Personalized Medicines.* The Human Genome Project, led by the National Human Genome Research Institute, part of the United States's National Institutes of Health (NIH), was an international research effort to sequence and map all the genes, together known as the *genome*, of members of *Homo sapiens.* It culminated in the completion of the full human genome sequence in April 2003. This gave the international scientific community the ability, for the first time, to read nature's complete genetic blueprint for building a human being. The challenge now is to discover the genetic basis for health and the pathology of human disease. In this respect, genome-based research will eventually enable medical science to develop highly effective diagnostic tools, to better understand the health needs of people on the basis of their individual genetic makeup, and to design new and highly effective treatments for disease. Individualized analysis on each person's genome will lead to a very powerful form of preventive medicine. It will be possible to diagnose risks of future illness on the basis of DNA analysis. Through understanding of the underlying "chemistry" of diseases such as diabetes, certain forms of cancer, heart disease, or schizophrenia, a whole new generation of drugs that are much more effective and precise than those available today will be found at the molecular level. As it takes more than a decade for a pharmaceutical company to conduct the clinical studies needed to win marketing approval from the FDA, most new drugs based on the completed genome are still perhaps 10–15 years in the future, although more than 350 biotech products, many based on genetic research, are currently in clinical trials, according to the Biotechnology Industry Organization.

3. *Biomarkers.* Biomarkers are measurable substances, such as proteins or metabolites, whose presence or concentration varies in response to a drug. Tools in the biomarker area will allow us to determine subsets of diseases, to predict individual patients' responses to treatment, to screen out people at high risk for an adverse event, and to monitor people during therapy so that they can be taken off a drug quickly if they are not responding to it. Thus, biomarkers will not allow discovery of new drugs per se but they can indirectly influence the R&D productivity by enabling physicians to prescribe drugs only to low-risk patients. Therefore, they will allow improvement of the risk/reward profile of a new drug.

4. *Informatics.* Many initiatives are underway to involve information technology (IT) in managing the enormous flow of data produced alongside drug development, such as combichem/HTS and genomics. NIH has identified bioinformatics as one new technological frontier in drug research, reengineering the clinical phase of drug development. Examples are:

- The project supported by seven big pharma companies to share internally developed laboratory methods to predict the safety of new treatments before they are tested in humans;
- The initiative-pooling experiences of big pharma companies, academic research, and biotech companies; and
- FDA's Critical Path to New Medicinal Products, also known as the Critical Path Initiative, aimed at networking clinical trials through integrative informatics.

Before there is any chance of such a network coming to fruition, with IT paving the way to a *global research* network, a nonproprietary attitude toward research will have to emerge.

5. *Virtual man.* Various academic institutes and bioinformatics firms are also building computer models of different organs and cells, with the ultimate aim of creating a "virtual man." Developing such a model will require a massive collaborative effort far exceeding that needed to complete the Human Genome Project. Nevertheless, predictive biosimulation is already playing a growing role in the R&D process, and we anticipate that, by 2020, virtual cells, organs and animals will be widely employed in pharmaceutical research.

6. *New Drug Approvals by Regulatory Bodies.* Side effects developed with approved drugs that have shown up after launch have resulted in a number of clamorous product withdrawals, such as Bayer's Lipobay (cerivastatin), Janssens's Propulsid (cisapride), Merck & Co.'s Vioxx (rofecoxib), Merrell-Dow's Seldane (terfendadine), Sankyo/Warner-Lambert's Rezulin/Noscal (troglitazone), and Wyeth's Fen-Phen (the nickname for a combination of fenfluramine and phentermine, each of which had been approved by the FDA to be taken separately to treat obesity).

These incidents have prompted the FDA to impose stricter requirements for drug assessment and approval. The observers of the doldrums in pharma innovation have raised the question as to whether new criteria applied are a root cause for the slump in new drug launches. Contrary to common belief, the FDA has *not* become more restrictive regarding New Drug Applications (NDA) approvals. Since 1991, the FDA has consistently approved 70–80% of all NDAs submitted. Although the "quality bar" has clearly been raised as clinical standards of care have improved, the FDA is certainly not as much of a bottleneck as critics claim. The reason is that prior to deciding on an NDA submission, the pharma industry itself looks very carefully at the approvability of a developmental drug.

The situation is likely to change, however, when *economic* criteria will also be taken into account in the drug approval process. The United Kingdom's National Institute for Health & Clinical Evidence (NICE), an independent organization responsible for providing national guidance on promoting good health and preventing and treating ill health, has pioneered the aspect of economy in drugs assessment, first elating to reimbursements by National Health. The German *Institut für Qualität, Wirtschaftlichkeit, Gesundheit* (IQWIQ) and about 40 more countries have similar institutions for implementing health technology assessment (HTA) methods.

Along the same lines, an EU initiative was started in 2006 to draw up common standards on HTA methods, which can increasingly make or break a new drug, even if it primarily addresses only the question of reimbursement by public health insurance. The hope is to bring greater rationality and speed to a process also known as the "fourth hurdle" for treatments to win approval, after safety, efficacy, and quality. Examples of drugs, respectively treatments, under investigation are:

- The use of Roche/Genentech's Avastin (bevacizumab) and Herceptin (trastuzumab) to treat early breast cancer, at a cost of close to $50,000 per patient per year.
- Pfizer's Exubera insulin inhalator treatment, which costs more than $2000 per year. NICE sees no justification for reimbursing a treatment cost three times higher just for the convenience of inhaling rather than injecting the insulin.
- IQWIG has ruled that there are no grounds for the clinical superiority of Lipitor claimed by Pfizer to justify a higher price than for generic "statins."

As things stand today, there are no shortcuts in drug development, regarding neither the overall development time—typically 12–15 years—nor the early identification of dropouts. The pharmaceutical industry will continue to have to contend with failure rates and timelines to bring products to market exceeding industry standards by far. Prominent scientists of the pharma industry, such as Malcolm McCoss of Merck & Co., John Lamattina of Pfizer, Robert Ruffolo of Wyeth, and Steven Paul of Lilly Research Labs, who were interviewed by *Chemical & Engineering News* [3], did not provide a clear answer to the key question, which stakeholders, shareholders, and patients alike, are most worried about: *"Will the new tools of pharma R&D allow the industry to also collect the 'high-hanging fruit and reverse the downward trend in new drug launches that has been continuing for almost 10 years now?"*

A recent report on the discovery of a new antibiotic by Merck & Co. scientists provides a good insight into the difficulties of developing a new drug:

The Merck team ... found the molecule, called *platensimycin*, by fishing it out of a library of 250,000 natural products. ... The molecule is produced by *Streptomyces platensis*, a bacterium isolated from a South African soil sample.

The formula, which contains (among other things) an aromatic and an oligocyclic moiety, is so complicated that a chemist who is not an expert in nomenclature cannot determine the chemical name. The structural formula is $C_{23}H_{25}O_7$.

The researchers custom-designed a screening assay to search for a FabF inhibitor in the extracts. They engineered a strain of *Staphylococcus aureus* to produce antisense RNA that gives them tight control of the amount of FabF produced. … Because the drug is rapidly cleared from the body, it requires continuous intravenous delivery.

In an effort to revert the negative trend in innovation and productivity and to boost innovation and R&D productivity, pharma is implementing *radical* changes in the R&D strategy, adopting a more entrepreneurial approach, simulating in-house the incentive-based, smaller-scale, science-based environment of biotech companies, tapping research conducted by other companies and academics and broadening their portfolios.

First, many pharma and biotech companies are downsizing their R&D divisions into smaller organizations. For example, GSK has created Centres of Excellence for Drug Discovery (CEDDs). These are small R&D units, containing 100–150 researchers. Each one is focusing on only one therapeutic area, such as oncology or respiratory disease. Individual CEDDs can start Phase II development without recourse to any other body. Pfizer is planning a similar move. Other companies are increasingly going to implement the new R&D model of the Networked Partnership (NP). For example, Lilly has set up an independent operation group called "Chorus" in its Indianapolis headquarters. Chorus focuses on handling all collaborations with its partners worldwide. It also has a similar operation group in its Shanghai R&D center, though this is much smaller in size.

Second, companies have diversified their research portfolios, shifting from their traditional reliance on chemical-based drugs to biological ones including vaccines.

Third, an increasing number of new drugs are licensed in from start-ups. Up to 50% of big pharma's R&D budget is being spent on the development of drugs licensed-in from virtual pharma companies, independent research organizations, and so on. This huge investment is due to the lack of innovation stemming from their own internal R&D efforts, requiring these companies to actively search for new products from third parties, as well as drive further mergers and acquisitions (M&A) activities. As new biotech-originated drug candidates for therapies targeting diseases such as obesity, central nervous system disorders, and cancer advance into clinical trials, the importance of partnerships between drug firms and small biotech specialists becomes more evident. It is expected that the top 10 big pharma companies will generate more than half of their sales from products derived from other organizations' R&D efforts in the future. Truly innovative research is also performed at governmental agencies, such as the NIH.

Fourth, companies are going to conduct broader and more extensive research on genomics and proteomics, including the development of disease/animal models, more effective diagnostic and imaging methods, and better biomarkers. Another priority is new techniques to detect, identify, and even reduce the toxicity of a drug compound. At present, more than 50% of drug candidates fail in clinical trials mainly because of their severe toxicity or side effects.

16.3 BUSINESS IMPACT

The combined effect of the patent cliff, low R&D productivity, and price pressure on drugs is expected to result in the smallest possible single digit revenue growth of the top 10 (see Table 11.2) pharmaceutical companies. Actually, financial analysts estimate their annual revenues increasing at a meager 1% CAGR, from $415 billion in 2009 to $440 billion in 2015. This calls also the marketing executives into action. They advocate breaking a dual taboo of the industry, namely selling only *originator drugs* in the *developed word*. The new strategy is to take advantage of the tremendous growth potential of the pharm-emerging countries. According to IMS Health, the pharma markets in Brazil, Russia, India, and China (BRIC countries) plus Turkey, South Korea, and Mexico are estimated to double from $93 billion in 2008 to $190 billion in 2013. These countries in the past had generated only a small part of big pharma's revenues. So far, only BMS and Eli Lilly had sales revenues of—slightly—more than 5% in the BRIC countries. Pfizer, which bashfully calls its generics business "Established Products" had the biggest revenues in absolute terms, $7.4 billion in 2009.

.... responding to structural pressures means a shift away from "white pills in western markets," with the proportion of traditionally core patent-protected, chemically based drugs, which are sold mainly in North America and western Europe, falling to just more than a quarter of total sales.

—Andrew Witty, CEO GlaxoSmithKline

In order to implement the new strategy, big pharma companies enter into alliances with local generic companies in the pharm-emerging countries. Thus, in 2010, Abbott acquired Nicholas Piramal's Generics business for $3.72 billion, corresponding to 8.7 times sales in 2010. The takeover catapults Abbott (or to be precise, its new "Established Products" division) to the number one position in generics in India. Further examples are Daichi-Sankyo's acquisition of Ranbaxy (number two in India), AstraZeneca and Torrent (India), GSK and Dr. Reddy's (India), Dong-A Pharmaceuticals (South Korea) and Aspen (South Africa), Pfizer with Aurobinda and Claris, and Sanofi-Aventis with Medley (India) and Kendrick (Mexico).

The most attractive class of pharmaceuticals will be branded generics, that is, generics carrying the name of a reputed big pharma company on the label. They give the patient the assurance of buying a safe drug. Because of the very competitive pricing for the APIs needed for these branded generics, fine chemical companies in "low-cost/high-skill" countries, particularly Chindia, will be the main beneficiaries of the forthcoming boom from this marketing strategy. In this context, "high skill" means compliance with the exacting standards requested.

> We will sell specialised generics in global markets but don't intend selling unbranded mass market generics.
>
> —Andrew Witty, CEO GlaxoSmithKline

CHAPTER 17

Generics

Pharmaceuticals are covered by two types of patents, product patents and process patents. The former refer to the therapeutic effects of the drug, for example, antihypertensive, the latter to the manufacturing process for the underlying API. After the expiration of the product patent, the inventor loses its marketing exclusivity. Depending on the legislation in individual countries and the speed of development, this typically takes place 18–20 years after filing and 8–12 years after launch. Third parties are then at liberty to sell copies of the drug. Two main restrictions have to be observed. The trade name of the originator product must not be used and, if a process patent is still valid, a noninfringing route for the synthesis of the active pharmaceutical ingredient (API) must be chosen.

In the narrower sense, the term *generic* is used for small molecule drugs. In such cases, the APIs are identical to those of the originator drug. They are predominantly produced by traditional chemical synthesis. If a generic company, for example, Teva, or the generic unit of an ethical pharma company, for example, Sandoz, registers a generic under a new name, it is called a *branded generic. Branded generics* are particularly popular in "pharm-emerging" countries, where patients shy away from medicines of doubtful origin and dubious quality. The generic versions of big molecules are called *biosimilars,* or *biogenerics*. Biosimilars have the same therapeutic effects as the originals, but, as the name indicates, the molecular structures of their complex and not exactly known so the APIs are not always 100% identical.

17.1 SMALL MOLECULE GENERICS

For three main reasons, the global market for generics is expected to expand at a rate of 7–7.5% per year, from $90 billion in 2009 to $135–150 billion by 2015:

Fine Chemicals: The Industry and the Business, Second Edition, by Peter Pollak
Copyright © 2011 John Wiley & Sons, Inc.

- Because of the "patent cliff" (see Chapter 16), the innovative pharmaceutical industry will lose $140 billion of revenues because of patent expirations by 2015. This will contribute about $35 billion to the $45–60 billion projected growth of the generics market.
- In the BRIC countries and other pharm-emerging countries, more and more people are using Western drugs, most of them in the form of generics or branded generics. This will contribute about $35 billion to the growth of the generics market.
- Governments in the Western hemisphere and Japan have embarked on programs to reduce health-care costs, a key element of which is more affordable medicines. The most important means to achieve that is to encourage the use of generics instead of patented drugs. Thus, health insurers only reimburse prices for generic versions, whenever they are available. The United States, where 70% of all prescriptions are for generics, is the most advanced in this respect, but other countries are following suit.

The $140 billion figure of proprietary drugs facing patent expiry by 2015 must not be mistaken for the business potential for embedded API-for-generics suppliers. Assuming a price drop of 75% for the formulated drug (see Fig. 12.3), a 50% slump in market share and that the API accounts for 15–20% of the price of the formulated drug, one arrives at a total API-for-generics market value of 2–2.5% or approximately $3 billion [$140 billion × (0.020–0.025)] for the year 2015. These incremental sales represent 30% of the present free market for API-for-generics (see Table 9.3). As about 75 products will lose patent protection, sales of about $40 million on a per product basis can be expected. The sales potential for the individual fine chemical company preparing the launch of a new product is further reduced by strong competition. The FDA gets about 10 Abbreviated New Drug Applications (ANDAs) each time a patent expires. The price collapse is a consequence of the loss of the exclusivity after patent expiry and the lower cost base of the generics industry; the slump in market share is caused by the lack of promotional support and the launch of new drugs in the same therapeutic category. As the generics market becomes more and more dominated by distributors and wholesalers (see the recent initiative of Wal-Mart below), the position of API suppliers within the whole supply chain is weakening. Teva and Sandoz are the top two generic companies. They have a sizable number of fine chemical manufacturing plants, are well integrated upstream, and intend to remain so. The other players outsource most of their API requirements. Therefore, in contrast to the situation for proprietary drugs, not much additional business can be expected in this area.

Teva of Israel is by far the largest generics company. With sales of $16.2 billion in 2009, it is more than twice as big as the world's number two, Sandoz ($7.5 billion). Shlomi Vanai, Teva's president and CEO, announced that he expects his company's revenues to double to $31 billion by 2015, maintaining a sales mix of

TABLE 17.1   Retail Prices of Originator Drugs and Generics in the United States, Germany, China, and India

USA		Germany		China		India	
Originator	Generic	Originator	Generic	Originator	Generic	Originator	Generic
Nexium, Astra Zeneca; *omeprazol*/60 tablets at 60 mg[a]							
$340	$320	$132	N/A	$160	N/A	N/A	$3½
Norvasc, Pfizer; *amlodipin*/100 tablets at 5 mg[a]							
$217	$25	$83	$15	$87	$60–80	N/A	$3
Plavix, Sanofi-Aventis, *clopidogrel*/100 tablets at 5 mg[a]							
$529	N/A	$390	$250	$305	$210	N/A	$16

[a] Drug prices from China and India are calculated from smaller packages (and strengths).
Sources: USA, www.drugstore.com; Germany, Rote Liste; China and India, drugstore prices.

70% generics and 30% branded drugs. Teva's growth is the latest reminder of the way generic drugs and the companies that make them are changing the global pharma landscape. It is no longer correct to think of generics manufacturers as scrappy little competitors nipping at the heels of big pharma.

The discount from which patients taking generics benefit varies very much depending on the market conditions. Whereas price reductions during the "180 days exclusivity" are small—about 10%—they can go up to 90% in the case of very competitive products, typically former blockbuster drugs, in pharm-emerging countries. In Table 17.1, the prices of Nexium, Norvasc, and Plavix, and their generic versions in four different countries are shown. The largest price difference between originator drugs and generics exist in the United States. Surprisingly, the prices for the originator drugs in Germany and China are in the same ballpark. By far the lowest prices for generics are found in India, but markets in other countries are also becoming more competitive.

The launch of Wal-Mart's "$4 Generics Program" sent shockwaves throughout the pharmaceutical industry, as well as its upstream suppliers and downstream distributors. Wal-Mart offers an approximately 1 month supply of 300 widely used prescription drugs at $4. An example in case is the well-known antidiabetes drug Glucophage (metformin HCl). The average pharmacy retail price for an approximately 1 month's supply of sixty 500 mg tablets of the generic is $43. In comparison, Wal-Mart's $4 price amounts to a staggering discount of over 90%! It can only achieve this by drastically shortening its supply chain, that is, sourcing bulk quantities of the formulated metformin HCl directly from the most competitive producers in China and India. For the API producer, the question is what this means for him. In the example chosen, the sixty 500 mg tablets correspond to a quantity of 30 g of drug substance, or a price of $133/kg. Assuming that the API accounts for 25% of the formulated drug's price, this leaves the producer with a target price of $33/kg.

Globally, more than 2,000 producers of API-for-generics exist. However, fewer than 10% meet highest standards in terms of GMP, IPR, and regulatory compliance and have a track record of serving regulated markets. China is now the global drug industry's primary choice for sourcing conventional, large volume APIs-for-generics, such as the various forms of penicillins. In the past 10 years, the Chinese pharmaceutical industry, which is a generics/API-for-generics industry, has grown at an annual rate of 30–40%. The portfolio of companies is very diverse. At the low end there are thousands of very small, "garage-type" firms; at the high end are several FDA-inspected firms with sales in the $300–1500 million range, such as Shijiazhuang Pharma (CSPC), with sales of $1500 million in 2009), Zhejiang Hisun Pharmaceutical in Taizhou, with $586 million, Harbin Pharmaceutical Group (HAYAO), and Zhejiang Hengdian. The industry's growth is also being accelerated by the influx of a large number of multinational service providers comprising, among others, sourcing offices of big pharma companies. For more modern API-for-generics, innovative Indian fine chemical companies are the frontrunners. An indicator of a country's portfolio diversity is the number of drug master files (DMFs) submitted. As shown in Table 17.2, India, holding a share of well over 50%, is by far the number one. Italy, which used to be the world leader until the early 2000s, is now relegated to position six.

Apart from the unfavorable cost situation (see Chapter 14), Europe's competitiveness is also in jeopardy because of the very exacting safety, health, and environment (SHE) regulations and unfavorable patent legislation, particularly the "180 days exclusivity" and the Roche Bolar provision. The latter prohibits the shipment of free samples of APIs for patented drugs. Furthermore, the European Supplementary Protection Certificate (ESPC) *de facto* prevents

TABLE 17.2 DMF Type II[a] Filings by Major Countries

Country	2008	2009
India	378	345
China	115	95
United States	73	65
Italy	54	55
Germany	21	14
France	10	7
Japan	20	21
Taiwan	17	18
Switzerland	10	8
Total	**698**	**628**

[a] A Type II DMF (US) covers drug substance, drug substance intermediate, and material used in their preparation, or drug product.

European-based API producers from being qualified as bulk sources before the effective marketing exclusivity date in their home country. In practice, this means that European producers are excluded from supplying API-for-generics for several years *after* patent expiration. In contrast to the rest of the world, the stockpiling of APIs for originator drugs in anticipation of patent expiration is also forbidden.

> In order to forgo the inconveniences, several European fine chemical companies have established subsidiaries in the "patent haven" of Malta. Examples are Crystalpharma (Spain), Medichem (both Spain), Dipharma (Italy), and Siegfried (Switzerland).

In order to curb imports of APIs that do not comply with Good Manufacturing Practice (GMP) standards, mainly from Asian producers, the EU issued Directive 2004/27/EC in Q3 2005. The legislation states that it is henceforth the responsibility of the holders of drug-marketing authorizations or pharmaceutical companies—and specifically of the relevant "qualified person" within these companies—to ensure that their APIs comply with GMP standards. A written and signed declaration certifying that each API supplied is made under GMP to the requirements of the international regulation ICH Q7 as described in the DMF filed with a country's health authority or with the European Directorate for the Quality of Medicines (EDQM) is requested henceforth. In order to improve the totally unsatisfactory implementation of the directives, the European Fine Chemicals Group (EFCG, a unit of European Chemical Industry Council [CEFIC]) asked the EU authorities to appoint officials for on-site inspections.

For the fine chemical industry, the short conclusion is: the main beneficiary of the double-digit demand growth for API-for-generics will be the Asian fine chemical industry.

17.2 BIOSIMILARS

Biosimilars appeared on the market much later than traditional generics. Not only were their proprietary "parents" invented about 100 years later than the small-molecule drugs, but financial, technical, and regulatory hurdles must also be overcome before commercial-scale production can start. These hurdles go back to the complex structure of the active ingredients. As stated at the beginning of the chapter, biosimilars have structures that are similar but not identical to those of the originals. In order to obtain approval for commercialization by the competent authorities, the manufacturers must show that their products are identical to the originator reference products with regard to master cell bank, as well as having the same efficacy, tolerance, and safety. In contrast with small molecule generics, extended clinical tests are required in order to provide the necessary data. Furthermore, the production of performing biosimilars requires very stringent production processes and the plants have to comply with exacting configurations.

Both the time needed and the development costs for biosimilars are much higher than for generics, namely 6 to 10 years and $80–120 million, respectively. In the most important prospective market—the United States—the regulatory framework has not yet been established. As part of the health-care reform approved in 2010, the pertinent legislation should pass Congress in due course. So far, a provision guaranteeing 12-year exclusivity to originator medicines before allowing biosimilars to enter the market has been approved. For the time being, only two biogenerics produced in living cell lines have got clearance for commercialization from both the European Medicines Agency (EMEA) *and* the FDA, both of which are produced by Sandoz: the growth hormone omnitrope, which is "biosimilar" to Pfizer's proprietary Genotropin, and binocrit, the biosimilar of Procrit (epoietin alpha). Up to 2009, the EU had approved seven biosimilars.

Sales of biosimilars in nonregulated markets are already in full swing. Whereas only hGh is close to launch in the United States, Europe has hGh, erythropoietin (EPO), and colony-stimulating factor (CSF), while China and India develop, produce, and market hGh, EPO, CSF, insulin, interferon, interleukin, and both mAb and non-mAb proteins. As shown in Table 17.3, India has a portfolio of 10 products offered by 8 different companies. Other countries developing a position in biosimilars include China, Indonesia, Iran, South Korea, and the Philippines.

TABLE 17.3   Biosimilars Made in India

Company	Biopharmaceutical	Biosimilar
Biocon	Insulin	*Insugen*
	Anti-EGFR	*BIOMAb*
	Insulin (Glargine)	*Basalog*
Cipla	Avastin	*tbd*
	Herceptin	*tbd*
Claris Lifesciences	Erythropoetin	*Epotin*
Dr Reddy's	G-CSF	*Grastim*
	Rituximab	*Reditux*
Ranbaxy	Neuropen	*Grafeel*
	Erythropoetin	*Ceriton*
Shatha Biotech	Erythropoetin	*Shanpoietin*
	Hepatitits B vaccine	*Shanvac*
	Stretokinase	*Shankinase*
Wockhardt	Insulin	*Wosulin*
	Hepatitis B Vaccine	*Biovac B*
	Erythropoetin	*Wepox*
Zydus Biogen	Erythropoetin	*Zyrop*

EGFR, epidermal growth factor receptor.

Within the $140 billion revenues of proprietary drugs, which are going to fall over the "patent cliff" by 2015 (see Chapter 16), are a number of biopharmaceuticals. Estimates for the size of the biosimilar opportunity vary in brand values between $55 and $75 billion in 2015. Recombinant protein-based first-generation blockbuster biopharmaceuticals, scheduled to come off patent in the next few years, include interferon, granulocyte-colony-stimulating factor (G-CSF), and human insulin. The patents for the second-generation products will start expiring by 2015. This comprises a number of blockbuster anticancer drugs, such as Roche/Genentech's Avastatin (bevacizumab), with sales $5.9 billlion in 2009), Herceptin (trastuzumab) with $5.0 billion, Rituxan (rituximab) with $5.8 billion, Erbitux (cetuximab) with $0.8 billon, and Tarceva (erlotinib) with $0.4 billion.

The potentially lucrative market for APIs for biosimilars has attracted the interest of leading pharmaceutical, generics, and fine chemical companies alike. The high R&D investment needed for the development of new biosimilars has led to a number of cooperation agreements. Notably, Lonza and Teva concluded a joint venture in 2009. By developing, manufacturing, and marketing "a number of affordable, efficacious, and safe biosimilars," they aim to become a leading global provider of biosimilars. Merck & Co. launched Merck BioVentures, also in 2009 and anticipates having five biosimilars in late-stage development by 2012. Collaboration agreements have also been concluded between Mylan and Biocon, Pfizer and Biocon (for insulin products), Ratiopharm and Mepha (for Filgastrim), and Stada/Hospira and Bioceuticals Arzneimittel (for EPO-zeta and Filgastrim).

CHAPTER 18

Other Life Science Industries

18.1 AGRO FINE CHEMICALS

The agrochemical industry has traditionally been viewed as a less attractive customer base for fine chemical companies than the pharmaceutical industry. First of all, the global agrochemicals market is much smaller than the one for pharmaceuticals. The 2009 revenues were $44 billion versus $835 billion, respectively. Moreover, both the facts that the end consumers, the farmers, are only willing to pay for pesticides if it makes economic sense and the large market share of generics—more than 70%—puts more pressure on the suppliers of the active ingredients. As the leading agrochemical companies are more backward-integrated than their pharma counterparts, there is less outsourcing. This is also supported by the fact that agrochemical companies have been very successful in product life-cycle extensions. One reason for the staying power of agrochemicals with the originator are "firewalls" put up by the authorities brought about by the agchem industry lobby to impede registration of generic formulations. A prime example is Monsanto's Roundup, which was the company's top-selling product until 10 years after patent expiry.

As a result of wider acceptance and a major R&D effort, sales of seeds and traits are growing more rapidly than sales of conventional pesticides. For both Monsanto and DuPont, they already exceed those of traditional agrochemicals. Obviously, the expanding use of pest-resistant, genetically modified (GM) crops (60% of the worldwide soybean acreage is already GM) will negatively impact the growth of demand for agro fine chemicals. On the other hand, there are a number of attractive aspects, too:

- The share of fine chemical companies producing agro fine chemicals is proportionally smaller.
- Although the world population is expected to grow only modestly, at a rate of less than 1% per year, from 7 billion in 2010 to 9 billion by 2040, the need for food is growing at a much higher rate, as less developed countries switch from plant- to animal-based nutrition. In order to produce

Fine Chemicals: The Industry and the Business, Second Edition. by Peter Pollak
Copyright © 2011 John Wiley & Sons, Inc.

1 kg of beef, 7 kg of corn are needed. Rapidly expanding production of biofuels also adds to the demand. In the United States, one-third of the maize production is already used for bio-ethanol production.

- The ongoing reduction of cultivable land calls for higher yielding crops.
- The agrochemical industry has been successful in shifting new product attrition to the relatively inexpensive research phase: Whereas two new products out of 140,000 investigated in research made it to the development phase and one to registration, more recent data are 140,000:1.3:1. Thus, if a fine chemicals company gets involved in the development phase, the risk of the project being abandoned is smaller than in the pharmaceutical industry. Currently, about 10 new products are approved every year, almost half as many as pharma.
- "Big Agro" is increasingly outsourcing. An example in case is Syngenta's €50 million (approximately $70 million) investment for expanding Saltigo's agro fine chemical manufacturing capacity at Leverkusen, Germany in 2010.
- The number of new chemical entity (NCE) introductions has remained fairly stable at around 10–15 per year over the past 20 years. This is a very impressive number compared with the new launches of the pharma industry, which spends much more on R&D. Also, the cost of bringing a new product to market has grown only moderately, from $152 million in 1995 to $184 million in 2000 and $258 million in 2005–2008 (for comparison, in pharma, it is more like $1.4 billion).
- A considerable rejuvenation of agro's product portfolio is taking place. Old, large-volume products with an unfavorable risk/reward profiles are being replaced by newer, safer, more specific, and more active ones. Approximately 40% of the older products have been removed from the market, leaving room for the introduction of new active ingredients. Whereas the old products were manufactured in volumes of tens of thousands of tons per year in dedicated plants, the requirement for the new molecules is in the range of several hundred tons per year. They also require more production steps, command higher unit prices, and are manufactured in multipurpose plants, which is the domain of the fine chemical industry. Typical examples for this transition are given in Table 18.1. With the Food Quality Protection Act calling for a reevaluation of the safety of more than 10,000 pesticides, the U.S. government is expediting this trend. In Europe, one-third of the well over 800 compounds that were approved in 1993 have not been accepted for re-registration.

Bayer CropScience [1] estimates that the global agrochemical market will grow 3% per year, to €56.3 billion (~$80 billion) in 2013 and 66 billion (~$95 billion) in 2018. The share of crop protection products will drop from 58% of the total in 2013 to 52% in 2018, while the share of seeds and traits will grow from 42% to 48%.

TABLE 18.1 Substitution of High-Volume/Low-Activity by Low-Volume/High-Activity Agrochemicals

Category	Conventional Products[a]		Modern Products	
	Group	Application Rate[c]	Group	Application Rate[c]
Herbicides	Phenoxies (older)	500–2500	Imidazolines[b]	50–100
	Cyclohexanediones (older)	200–500	Sulfonlyureas[b]	10–100
		200–3000	Phenoxies (newer)[a]	100–200
	Triazines		Cyclohexanediones (newer)[b]	50–100
Insecticides	Organochlorines	100–10,000	Pyrethroids[a]	5–250
	Organophosphates	200–5000	Neonicotinoids[b]	5–200
Fungicides	Dithiocarbamate	250–3500	Triazoles[a]	5–250
	Morpholines	750–1000	Strobilurins[b]	50–200

[a] Key patents expired prior to 2005.
[b] Key patents expiring after 2005.
[c] Grams per hectare.
Source: Agranova.

18.2 ANIMAL HEALTH INDUSTRY

As reported in Section 11.3, animal health, with global sales of about $20 billion per year, is the smallest segment of the life science industry. Three developments will prove beneficial for the fine chemicals industry. First, the *farm animal* sector, representing 60% of revenues, will benefit from rapidly expanding industrialization of animal production as a prerequisite for satisfying humanity's rapidly increasing meat consumption. Second, the *companion animal* sector (40%) will profit and form the increasing share of the population, which can afford to own and care for pets. The intrinsic growth of the industry due to these drivers is expected to be 5% per year for the next 5 years. Third, as more and more animal health companies become independent, they are no longer dependent on their (former) parent pharma companies for the supply of the drug substances. Hence, they will turn to third parties both for procurement of active pharmaceutical ingredients (APIs) and custom manufacturing services of their veterinary drugs' active ingredients. An example in case is the newly formed number one animal health company formed in 2010 by the merger of the world's former number one, Intervet of the United States, and the number three, France's Mérial. Sooner or later, the new firm will unlock completely from its former parent companies, Schering-Plough and Sanofi-Aventis, respectively. With a market share of close to 30%, it will have a dominant position. Moreover, the current trend of pharma to scale down in-house fine chemical manufacturing will also prove beneficial for fine chemical companies serving the animal health market. Thus, when Evonik-Degussa acquired Eli Lilly's Tippecanoe Laboratories in Indianapolis in 2009, it got hold also of a sizable portfolio of veterinary drugs of Lilly's animal health subsidiary, Elanco. The site has fermentation capacity for, among other things, the polyether antibiotic Rumensin (monensin) and the macrolide anti-antibiotic Tylan (tylosin). In an adjacent fine chemical plant, the acetyl isovaleryl derivative Micotil (tilmicosin) is manufactured. The growth pro-moter Somatropin is another high-molecular-weight (HMW) molecule, where a substantial merchant market is developing. On the other hand, the animal health industry is not very innovative. Therefore, there will be limited business opportunities for custom manufacture of new molecular entities (NMEs).

CHAPTER 19

Contract Research Organizations

Contract research organizations (CROs) provide tailored product development services to the chemical and life science industries, allowing their customers to manage product development more efficiently and cost-effectively. This is a rich reservoir of opportunities for a business, which already generates revenues of about $25 billion per year, equivalent to one-third of global pharma's R&D spending. The market is expected to grow at a rate of about 10–15% to reach approximately $40–50 billion in 2015. CROs could be involved in up to 50% of drug development projects, both at the preclinical and the clinical stage, by 2015. It must be noted, however, that the lion's share, about 75%, of CRO revenues are generated from "patient" (medicinal) research and only 25% from "product" (chemical) research. The modern tools of pharmaceutical research are creating an enormous number of new lead compounds (see Section 15.2), which are funneled into the new product pipelines. Furthermore, there is a trend toward more outsourcing of R&D activities in the life science industry. According to a recent study of Morgan Stanley [2], downsizing internal research and partially reinvesting in later-stage licensing deals "could triple the number of new drugs reaching the market every year from 2014, generate higher near-term EPS and lead to significant increase in NPV." Not surprisingly, therefore, the new outsourcing trend of big pharma companies is to give up even mission-critical activities and depend solely on a CRO partner. These new "strategic service partnerships" may involve also the transfer of pertinent technical staff to the CRO.

Eli Lilly, for instance, has concluded strategic service partnerships with patient CRO Covance for toxicology studies, Fisher Clinical for dose and CMC development, Quintiles for clinical monitoring and i3 for clinical data management, and with product CRO Jubilant Life Sciences for chemistry and biology.

The attractiveness of offshoring CRO activities depends on the particular task to be performed. As illustrated in Figure 19.1, custom synthesis, where the competitive advantage of Indian CROs comes to full fruition, is part of the "high attractiveness" category. As a result, the offering of product CRO services is an attractive business proposition.

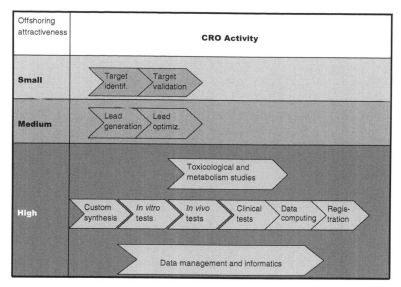

Figure 19.1 Offshoring attractiveness of CRO activities.

A note of caution is necessary for companies seeking to enter contract research in Western countries and Japan. Because of the low salaries of Chinese and Indian scientists and their longer work hours, the cost advantage of Asian competitors is even more dramatic than in other segments of the fine chemical industry. Whereas the prime cost of Indian custom manufacturers is about 40% of their Western competitors, it is only 20% for contract research. Attracted by full-time equivalents (FTEs) of $60,000 in the East versus $200,000 in the West, big pharma in particular is gravitating to these firms at the expense of their traditional U.S.- and EU-based CROs. In order to address this issue, Western companies are establishing footholds in Chindia.

> Flamma, a small Italian fine chemical company specializing in amino acid deri-
> vates, leased laboratory space at Shanghai's Zhangjiang Hi-Tech Park, the
> so-called "Medical Valley" of China. The R&D group comprises ten chemists
> with a bachelor's or master's degree, headed by one PhD, a former U.S. expatri-
> ate. The cost of the chemists amounts to less than 30% of their European peers.

In the *United States*, there is a business potential with small and medium-sized, chemistry-poor biotech companies. They often are betting on just one compound and are not willing to run the risk of a long-distance relationship that could go wrong. A number of U.S. and Canadian CROs located on the West Coast, where there is also a high concentration of virtual pharmaceutical companies, are specifically targeting this niche market.

In *China*, fully equipped, "ready-for-use" technology centers have been and are being built. In *India*, Jubilant Organosys, after having opened an impressive new research facility, now operates three laboratories staffed with nearly 250 scientists at its Noida site, 40 min from New Delhi. Hyderabad's GVK Bio is constructing a new R&D center that includes laboratories for 150 scientists. Hikal's Acoris subsidiary has built a state-of-the-art, ecosystem-friendly research center for contract research at Pune. Once it is fully operational, it will house 250 scientists. Hikal thus will join the ranks of integrated contract research/manufacturing companies.

India is an R&D "hotspot" defined as a place where (1) companies are able to tap into existing scientific and technical expertise networks, (2) there are good links to academic research facilities, (3) the environment supports innovation, and (4) it is easy to commercialize. The costs of pharmaceutical innovation in India are estimated to be as low as one-seventh of their levels in Europe.

—Economist Intelligence

Challenges for product CROs are the forthcoming competition from full service CROs, such as Charles River, Covance, Kendle, Organosys and Quintiles, and the increasing importance of generics. As a consequence of the growing importance of API-for-generics, a modification of the CRO business model will be necessary. Instead of customer-sponsored research, the firms will have to develop novel manufacturing processes for API-for-generics out of their own financial resources. A payback will materialize once the processes are out-licensed.

CHAPTER 20

Conclusion: Who Is Fittest for the Future?

Two partially interdependent conditions are creating both unprecedented opportunities and unprecedented threats for the fine chemical industry. First, for a number of reasons, the golden age of the industry's major customer base, pharma, has come to an end. In order to address the challenges, it is implementing far-reaching *restructuring* programs [3]. As a result, it will rely more on outsourcing R&D and manufacturing activities, without, however, moving to a totally virtual mode of operations for R&D or supply chain. Divestments of R&D and manufacturing units are also part of the program. Therefore, opportunities arise for acquiring plants and/or taking over entire product portfolios and new business prospects for API-for-generics in Asia (see Table 20.1).

Second, *globalization* is impacting the life science industry, too. The markets for Western-type medicines in the "pharm-emerging" countries will enjoy a rapid growth. In order to benefit from the booming markets, Western big pharma companies enter into alliances with local pharma companies. The resulting demand for APIs, mainly API-for-generics, will almost completely be covered by regional fine chemical companies. With their "high-skill/low-cost" advantage, they will also more and more become suppliers to the Western life science industry, to the detriment of the Western fine chemical industry. Its optimistic growth vision [4] has not been fulfilled, and it will be relegated to small volumes of late-stage development drugs (see Table 20.2).

The questions that need to be asked, therefore, are: "What does it take a fine chemical company to survive in this harsh business environment?" and "What are the important differentiators distinguishing the winners?" There are many criteria for assessing a fine chemical company. When auditing potential suppliers, GSK, for example uses a checklist with 22 elements, such as *Quality Systems, Tool Box, Performance Measures (Scorecard), Supply Contingencies, Continuous Improvement*, and *Price Reduction Performance* (see also Appendix A.6). Syngenta looks for 10 supplier attributes, namely

---

TABLE 20.1 Pharma Restructuring

Restructuring	Consequences for the Fine Chemical Industry
"Patent Cliff": $140 billion of originator drug sales affected by patent expiries	Many new business opportunities for API-for-generics
Few new drug approvals (no change)	Few opportunities for custom manufacturing PFCs for new drugs
More outsourcing	Opportunities for CM of PFCs for mature drugs
Plant divestments	Opportunities for acquisition of plants and/or taking over of entire product lines
Foray in pharm-emerging countries	Business opportunities for domestic companies

PFCs, pharmaceutical fine chemicals.

TABLE 20.2 Globalization

EU + US		Asia (Chindia)
	Purchasing	
Relational	←————————————→	Transactional
	Drug development stage	
Late	←————————————→	Early
	Product type	
Exclusives	←————————————→	API-for-generics
	Volume requirements	
Small	←————————————→	Large
	Type of molecules	
Small (traditional)	←————————————→	Big (biosimilars!)

Strategic Alignment, Innovation Potential, People, Value Potential, Corporate Principles & Risk, Organisational Fit, Trust, Supply Chain Capability, Technical Capability, and *Supply Fit.* Many of these criteria are qualitative and leave room for personal judgments by the auditor. If this were not the case, all customers would choose the same supplier.

Ultimately, however, there is one overriding metric for the qualification of an individual fine chemical company:

a track record of successfully completed projects with major customers

The author has taken a bold approach and selected just three key performance indicators (KPIs) required to get to such an enviable "preferred supplier status," namely:

1. *Globalization, in particular a presence in Asia*

Apart from pharma restructuring, globalization is the single most important trend affecting the fine chemicals industry. If a company does not take advantage from the "high-skill/low-cost" opportunity advantage in Chindia *and* the rapidly expanding domestic demand in this 2.3 billion people region, its long-term survival is at risk.

Examples:

"High skill": The impressive pipeline of 10 biosimilars with Indian life science companies (see Section 17.2)

"Low cost"/contract research organizations (CROs): The full time equivalent (FTE) is the "currency" for charging services by CROs. An FTE costs about $50,000 per year in India, one quarter of the price tag in Europe or the United States—and an Indian chemist works 30% more hours per year.

"*Low cost*"/contract manufacturing organization (CMO): Operational costs for a current Good Manufacturing Practice (cGMP) fine chemical plant equipped with $4\,m^3$ reaction vessels are $8/m^3$/h for an Indian custom manufacturing [CM]), as compared with $40 for a Western CM and $80 for a "big pharma" unit (see Table 14.3)!

In a broader sense, the above-average growth scenario applies to all of the pharm-emerging countries (Brazil, China, India, Indonesia, Mexico, Russia, South Korea, and Turkey). They are expected to account for as much as 20% of global pharmaceutical sales by 2020—a 60% increase since 2004. As described in Chapter 14, the U.S. fine chemical industry already has become insignificant. In addition, the playing field for the hitherto dominant Western fine chemical companies is shrinking—and might even disappear for those companies which do not have at least co-operations with Asian peers, the top tier of which have built plants and adopted business processes up to Western standards.

If the double-digit growth of the Chindian fine chemical industry continues, a horror vision could become true. Western fully EMEA-, FDA-, REACH-, and cGMP-compliant plants would stand idle, and the best trained people would be unemployed. …

2. *Ownership and size*

In this context, size is meant in absolute terms for a stand-alone company, or in relation to the size of the parent in the case of a unit of a large company. In general terms, a diversified, globally acting Chemical Conglomerate is not particularly suitable for a service-oriented custom manufacturer. Among stand-alone firms, a privately or family-owned midsized company with sales in the range of $100–250 million has the best prerequisites [5]. It is big enough to be able to offer an attractive technology/service portfolio and at the same time small enough to provide flexibility and short response times. Flexibility has broad implications in terms of operational, financial, and legal business aspects. Examples are critical aspects of customer orientation, such as the introduction of new analytical methods, the timing of production campaigns, the acceptance of nonconventional payment terms and suppleness with regard to the terms of supply contracts. These competences are particularly important in custom manufacturing (see Section 12.2.1). Not being under the permanent pressure from Wall Street is an additional advantage. Whereas the time horizon of the publicly owned company is the next quarter, it is the next generation for the former. Family-owned companies are reluctant to lay off employees and shut down plants during troughs in the business cycle. Therefore, the know-how is preserved, and both the human and physical assets are ready to seize new opportunities as soon as they come up.

If fine chemicals does not comprise a large unit within a diversified company, it cannot be run successfully. It simply does not get the necessary attention from management and is at a disadvantage in terms of resource allocation. Therefore, fine chemicals must be a core competency of a company. Jack Welsh, the former CEO of GE stated that a company should only be engaged in activities, where it can hold the top or second place position. This rule applies also to individual divisions of diversified companies. As the $100–250 million bracket applies also to this case, the parent company's sales should not exceed the approximate range of $300 million–$1 billion.

As a matter of fact, tracking back past performance of the industry, it has become evident that midsized, family-owned fine chemical companies outperform small fine chemical business units of large companies.

An example in case is Saltigo. The company was founded as a spin-off from Lanxess in April 2006. Lanxess in turn had been spun-off from Bayer in 2005. With both predecessors, the fine chemical business underperformed. The chairman of Lanxess, Axel C. Heitmann, described the situation as follows: "Fine chemicals was a business that was essentially neglected and sidelined for the past decade and had produced hundreds of millions of dollars of losses. ... Saltigo managed a turnaround and became profitable in the first year as a "pure player."

An opposite development has taken place with fine chemical divisions of large, diversified chemical companies, such as Bayer, Clariant, Dow, Eastman, Honeywell, Rhodia, and Solutia: they do no longer exist (see also Section 14.1)!

3. Differentiating technologies, especially biotechnology

Within *conventional chemistry*, there are a number of niche reactions offering better growth potential than their widely used conventional counterparts. In terms of chemicals used, this applies for instance to the handling of hazardous gases, such as cyanogen chloride, diazomethane, fluorine, hydrogen cyanide/fluoride/sulfide, mercaptans, nitrous oxides, ozone, and phosgene. The risk-adverse pharma industry prefers to avoid these chemicals. On the downside, the entry barriers are low. Ozone and phosgene reactors, for example, can be bought "off the shelf" and their installation is usually not on the critical path of a project. Solids requiring special handling precautions include antibiotics, controlled substances, and steroids. Out-of-the-box technologies requiring special equipment comprise microreactors, predestined for energetic reactions, simulated moving bed (SMB) technology, used, for example, for racemate separations, and high containment facilities, used for the synthesis of high-potency active ingredients (HPAIs). Within the latter, the syntheses of antibody–drug conjugates (ADCs) constitute a particularly appealing niche. If one includes API-for-biopharmaceuticals, the potential of the HPAI market could be $5 billion by 2015. Thus, it exceeds by far the one of the other niche technologies. Despite the high growth rates, their market will stay within a few hundred million dollars per year in the foreseeable future.

With *biotechnology*, the situation is totally different. Processes are demanding, especially the mammalian cell culture technology for producing "big molecules," that is, biopharmaceuticals, is mastered only by a handful of companies on an industrial scale. The market potential is in the order of billions of dollars. On the downside, the entry barriers are very high—among others, demanding technologies, strong regulatory constraints, and high capital intensity, both in fixed assets and working capital.

A broadening of the range of services, such as combining primary and secondary manufacturing or offering a combination of CRO and CMO services (or CRAMS [contract research and manufacturing], to use the Indian term), with the idea of becoming an all-in-one partner of life science companies is *not* considered a key success factor. It is a great challenge for a midsized company to master different competences at the same time, while customers adopt a functional approach in their supply strategy and select the most qualified, specialized supplier for each competency. As a *deus ex machina* for solving the fine chemical industry's problems, consolidation has been the mantra of many consultants—but it still has to take place.

It is much harder to be in this business today than it was five or ten years ago.

—Guy Villax, CEO Hovione

CITED PUBLICATIONS

1. N. Alperowicz, *Bayer boosted by new agchems; ponders environmental science,* *Chemical Week*, September 28/October 5, 2009, p. 16.
2. Morgan Stanley, *Pharmaceuticals—Exit Research and Create Value*, January 20, 2010.
3. Maureen Rouhi et al. Pharma's road ahead, *Chemical & Engineering News*, June 19, 2006, pp. 24–99.
4. A. Liveris, Has the outsourcing phenomenon fulfilled its promise, *Specialty Chemicals Magazine*, May 2003, pp. 28–30.
5. P. Pollak, Family owned businesses fare better, *ChimicaOggi/Chemistry Today* 26 (**6**), 2 (2008); G. Villax, Family owned businesses, *ChimicaOggi/Chemistry Today* 26 (**6**), 8–10 (2008).

▰▰▰ ABBREVIATIONS

ADC	Antibody drug conjugate
ADMET	Absorption, distribution, metabolism, excretion, and toxicology
a.i.	Active ingredient
ANDA	Abbreviated New Drug Application
API	Active pharmaceutical ingredient
API-for-generics	Active pharmaceutical ingredients for generics
BRIC	Brazil, Russia, India, and China
b.u.	Business unit
CDA	Confidentiality disclosure agreement
CAGR	Compound annual growth rate
CDER	Center for Drug Evaluation and Research (department of FDA)
Chindia	China and India
CHO	Chinese hamster ovary
CM	Contract (*also* custom) manufacturing
CMO	Contract manufacturing organization
COGS	Cost of goods sold
combichem	Combinatorial chemistry
CPhI	Chemical and Pharmaceutical Ingredients (exhibition)
CRAM	Contract research and manufacturing
CRO	Contract research organization

Fine Chemicals: The Industry and the Business, Second Edition, by Peter Pollak
Copyright © 2011 John Wiley & Sons, Inc.

DP	Drug product
DQ_E	Design qualification estimate
DS	Drug substance
EBIT	Earnings before interest and taxes
EBITDA	Earnings before interests, taxes, amortization, and depreciation
ECHA	European Chemical Agency
EFCG	European Fine Chemicals Group
EMAS	European Union Eco-Management and Audit Scheme
EMEA	European Medicines Agency
EVA	Economic value added
FDA	U.S. Food and Drug Administration
F&F	Flavors and fragrances
FTE	Full-time equivalent
GM	Genetically modified; general manager
GMP	Good Manufacturing Practice
HMO	Health management (*also* maintenance) organization(s)
HMW	High molecular weight
HAPI	Highly active pharmaceutical ingredient(s)
HPAI	High-potency active ingredient(s)
HPLC	High-pressure liquid chromatography
HTA	Health technology assessment
HTS	High-throughput screening
IP	Intellectual property
IPR	Intellectual property rights
IQ	Installation qualification
IQWiG	Institut für Qualität und Wirtschaftlichkeit im Gesundheitswesen
IT	Information technology
JV	Joint venture
KPI	Key performance indicator(s)

LC	Liquid crystal
LCD	Liquid crystal display
LMW	Low molecular weight
mAbs	Monoclonal antibodies
M&S	Marketing and Sales
MHRA	Medicines and Healthcare Products Regulatory Agency (United Kingdom)
MP	Multipurpose
MRT	Microreactor technology
MW	Molecular weight
NAI	No action indicated
NBE	New biological entity
NCE	New chemical entity
NDA	New drug approval (*or* applications)
NIH	National Institutes of Health
OEE	Overall equipment effectiveness
OLED	Organic light-emitting diode
OQ	Operational qualification
OTC	Over-the-counter (drugs)
PAT	Process analytical technology
PBB	Peptide building block
P/E	Price/earning ratio
PFC	Pharmaceutical fine chemical
P&L	Profit and loss
PoC	Proof of concept
POLED	Polymer light-emitting diode
PQ	Performance qualification
QA	Quality assurance
QC	Quality control
RDA	Recommended daily (*or*) dietary allowance

REACH	Registration, evaluation, authorization, and restriction of chemicals
RFI	Request for information
RFP	Request for product
R&D	Research and development
RONOA	Return on net operating assets
ROS	Return on sales
RS	Requirement specification
SEZ	Special economic zones
SFC	Supercritical fluid chromatography
SFDA	State Food and Drug Administration (China)
SHE	Safety, health, and environment
SMB	Simulated moving bed (chromatography)
SOP	Standard operation procedures
TRIPS	Trade-related aspects of intellectual property rights
USP	Unique selling proposition
UHTS	Ultra-high-throughput screening
VTO	Volume time output

APPENDICES

Information Sources/Life Sciences

Company, Organization	Website	Contents
PHARMA		
	Databanks and Consultants	
Arthur D Little International, Inc., Brussels	www.arthurdlittle.com	Market research, strategy consulting, "round robin" financial data of major FC companies
Biotech companies	www.business.com/directory/ pharmaceuticals_and_biotechnology/ biotechnology	List of biotechnology companies
CIPSLINE, Prous Science, Barcelona, Spain	www.prous.com	Developmental drugs database
eKnowledgeBase/Engel Publishing Partners West Trenton, NJ, USA	www.pharmalive.com	Online development drug database with more than 20,000 listings
ePocrates RX™	www.epocrates.com	Clinical drug database
Drug Database, Reuters Health	www.reutershealth.com	Brand names, dosage, therapeutic category, and more on drugs, USA
Drugstores, USA	www.drugstore.com	Drug retail prices in the USA
F-D-C Reports Inc., Chevy Case, MD, USA	www.fdcreports.com	Pharma information
IMS Health, USA	www.imshealth.com	Global statistics on drug sales
		Pioneer databank on developmental drugs
Investigational Drugs Database, Current Drugs Ltd., London	www.current-drugs.com	Developmental drugs database (with chemical structures)
Pharmaceutical Research & Manufacturers of America	www.pharma.org	Pharma industry sponsored, patient oriented info's on existing and new drugs
Pharmaprojects, PJB Publications, Surrey, UK	www.pjbpubs.com	Developmental drugs database

Pioneer, see IMS Health

Reuters Health (see: drug database) www.reutershealth.com Brand name, dosage, therapeutic category, etc.

SciFinder www.cas.org/SCIFINDER Scientific lit./patents/new technologies

Technology Catalysts www.technology-catalysts.com Biopharmaceuticals Information & Consulting Services

AGRO

Agranova www.agranova.co.uk AgChem New Compound Review (NCR);

www.agranova.co.uk/agrall1.asp Agranova Alliance, a consortium of agchem info providers

Agvet Reports, PJB Publications, Richmond, Surrey UK www.pjbpubs.com News on the agrochemical industry

Becker & Associates, Paris, France www.beckerdata.com Agrochemicals database

ENIGMA Marketing Research, Goostrey, UK emrgoostrey@msn.com Agro market research

Philips McDougall—AgriService www.phillipsmcdougall.com

Alan Wood, London, UK www.alanwood.net Pesticide Compendium

ANIMAL HEALTH

Vetnosis, Edinburgh, UK www.woodmacresearch.com Animal health market research

Magazines

Chemical & Engineering News, ACS, Edison, NJ, USA //pubs.acsorg/cen General chemistry, emphasis on R&D

Chemical Week, NY, USA www.chemweek.com General chemistry

Chimica Oggi, Chemistry Today, Milan, Italy www.teconscience.com General, fine, and specialty chemistry

TABLE *(Continued)*

Company, Organization	Website	Contents
ICIS Chemical Business (formerly *Chemical Market Reporter*)	www.icis.com	Leading U.S. monthly chemical business review (free monthly newsletter also available)
MedAd News, Newtown, PA, USA	medadnews.com	List of world's 200 best-selling prescription drugs (May issue)
Pharmaceutical Technology/Advanstar Comm. Iselin, NJ, USA	www.pharmaceutical-technology.com	Articles and press releases on the pharma and FC industries; investment projects
Scrip Magazine/Informa Healthcare, London	www.pjbpubs.com	"Yellow pages" = pharma newsletter
Speciality Chemicals Magazine/dmg world media Redhill, Surrey, UK	www.specchemonline.com	Articles on speciality and fine chemicals

Training Courses

ISPE, International Society for Pharmaceutical Engineering Tampa FL, USA	www.ispe.org	Computer control, containment , GMP, pharma-engineering, , etc.
Scientific Update, Mayfield, East Sussex, UK	www.scientificupdate.co.uk	Chemical process development and scale-up in organic chemistry

Directories

Chemical Buyer's Guide	www.buyersguide.com	
Buyer's Guide—China	wwwchinachemnet.com	
ICIS Search	www.icis.com/Search	Provides free offers for chemicals

Exhibitions

BIO International Convention	www.bio2010.com	Leading global biotechnology event
CPhI, Convention on Pharmaceutical Ingredients	www.cphi.com	Major fine chemicals trade show, annually
Informex	www.informex.com	Fine, Specialty & Custom Chemistry

Governmental Organizations

European Agency for the Eval. of Medicinal Products	www.emea.euorpe.eu	European equivalent of FDA
European Chemicals Agency (ECHA)	www.echa.europe.eu	Official body for REACH
FDA, Food and Drug Administration, USA	www.fda.gov	Regulations and studies [Drug Eval. and Res.
	fda.gov/cder/da/da.htm	FDA new drug approval list (CDER[a])
		Orange book listing drug patents
SFDA	www.sfda.gov.cn	Chinese equivalent of FDA

[a] Center for Drug Evaluation and Research (department of FDA).

Checklist for New Product Evaluation

Product name, structure		Raw material, advanced intermediate, API
Market Information		
End product		Brand name
Therapeutic category		Novelty, me too
Competitive products		Product, company advantages/disadvantages
Status		Phase I, II, or III
Year of launch		
Customer(s)		Big, mid, or virtual pharma, existing customer; previous business
Competition for **A**		Who else has been contacted; Competitive advantages/disadvantages
Total demand potential for **A** additional information		**A**'s expected market share at maturity
Project Status		
Negotiations		Secrecy agreement, R&D agreement, order for trial quantity supply agreement
R&D work		Man-months spent vs., man-months needed
		Total R&D budget, ratio R&D cost vs. sales
Samples		Samples sent/approved

Time/volume Sales (mt)	2009	2010	2011	2012	2013	2014	Past/expected sales volume	
Sales price							A's offer, customer expectation	
Turnover								
Raw materials							Key raw materials, availability	
Plant/ investments	*Plant*	mp	special	*Investment*	pp	works	Fit with existing plant, adaptation investments	
	existing			add.				
	new			new				
Environment							Safety/toxicity; waste volume/disposal	
Conclusion								
Pro's and con's							Profit expectations/ fit with in-house technologies/ chances of success	

Project Schedule, Custom Manufacturing Project

Main Activities	Timing	Management

• Preparation and submission of a preliminary offer, based on desk evaluation
• *(After acceptance by customer)* decision on go/no go
• Outline of project task
• Conclusion of CDA and technical discussion with customer
• Formation of the project organization. Nomination of project champion, leader, and team members
• Fixing of the research program with objectives, milestones, resources, budget, etc.
Note: This program will constitute a central part of the offer
• Sample preparation and submission to customer
• Preparation of a detailed offer with the following sections:
  1. Executive Summary
  2. Project history (first contact, preliminary agreements, etc.)
  3. Technical part: R&D program; production schedule
  4. Commercial part: detailed offer for laboratory (basis: FTE), pilot plant phase, and supply of industrial quantities; assumptions; investments for plant adaptation.
  5. Project organization and communication channels; reporting
  6. Timelines
  7. Blueprints of the pilot plant and production trains
• *(After acceptance by customer)* ↘
	1–2 months	During R&D phases monitoring by "New Product Committee," is project "within budget and. timelines," are milestones completed, have customer expectations changed, etc.

<div align="center">Project start</div>• Technology transfer customer → supplier • Process research; route and sequence selection, analytical method development • Process development: identification, exploration, and optimization of critical parameters • Safety and ecology data • Conclusion of laboratory phase ↘	6–9 months	
<div align="center">Transfer to pilot plant</div>• Proof of concept of the laboratory process • HAZOP • Validation of analytical methods • Definition of final specifications • Production and supply of trial batches • Regulatory submissions • Conclusion of pilot plant phase ↘	6–9 months	
<div align="center">↘ Transfer to industrial plant</div>• Confirmation/revision of the offer for industrial scale supplies • Conclusion of supply contract • Plant adaptation • Production scheduling, operator instruction • Start R&D work on a second-generation process • Master batch records • Optimization of cycle times • Start of industrial-scale production • Administration of the supply contract (updating of supply schedules, price adjustments, market intelligence)	Several years	Current business

Company Scorecard

	Activity	Target
	Financials	
1	Development of Economic Value Added	Replacement cost of capital
2	Cost/m^3 × hour	<$25
3	Reduction of working capital	Inventory turnover minimun 4× per year
	Manufacturing	
5	Continuous cost improvement	COGS reduced by 10–20% each time the yearly production volume doubles
6	Number of reworks/total batches	<1%
7	% on time shipments	>99%
8	Number of quality complaints	To be determined
	R&D	
9	New product[a] sales as % of total sales [a]introduced in the past 5 years	>20
10	Successfully completed projects as % of all projects initiated in R&D	>25%
11	% of successful scale-up's laboratory → pilot plant	To be determined
	Marketing and Sales	
12	Fulfilment of budget/business plan	Sales and profit growth > benchmark
13	Number of tier 1 customers	Frame supply contract agreed upon
14	New business	min. 2 phone calls per week min. ? new projects,/? from new customer visits

Fine Chemicals: The Industry and the Business, Second Edition, by Peter Pollak
Copyright © 2011 John Wiley & Sons, Inc.

TABLE (*Continued*)

Activity	Target
Engineering	
15 % of projects "on time/on budget[a]	To be determined
16 Installation cost/m³ reactor volume	<$500,000/m³
17 Plant availability	To be determined
Safety, Health, and Environment	
17 Yearly reduction of total emissions	To be determined
18 Number of lost hours due to accidents	To be determined
19 Hours of formation/ employee × year	To be determined
Human Resources	
20 Fluctuation rate	<5% per year
21 Absenteeism	<10 (lost hours × 10⁶/Σ hours)
22 % of revenues spent on formation	To be determined

[a] COGS, cost of goods sold.

Job Description for Business Development Manager

Primary Responsibilities

- Sustains the growth of company's business by
  - Establishing new business relationships within the Life Science industry
  - Acquiring ideas for new fine chemicals/API-for-generics/custom manufacturing arrangements
  - Evaluating the business ideas according to the company criteria
  - Carrying out the supporting market studies
  - Defining the scope of the project
  - Assuming the "project champion" function for the realization
  - Calculating sales prices in accordance with company rules for profitability
  - Negotiating and concluding confidentiality agreements, R&D agreements, supply contracts for trial and industrial quantities
- Project Management
  - Chairs the "New Product Committee"
- Customer Relations Management
  - Implements effective key account management
  - Monitors and updates Quality Performance Database (scorecard)
  - Identifies decision makers at the major accounts
  - Cultivates the relationship with the major accounts
  - Ensures flawless flow of communication
  - Assures proper administration of supply contracts
- Market Intelligence in the following areas
  - Life science industry in general
  - New product pipelines of the major potential and existing customers
  - Competitor intelligence
- Advertising and Promotion
  - Edits company literature, product profiles, company website
  - Organizes participation at trade shows
  - Organizes customer events

Fine Chemicals: The Industry and the Business, Second Edition, by Peter Pollak
Copyright © 2011 John Wiley & Sons, Inc.

TABLE (*Continued*)

Support Functions

- Strategy/Corporate Development/Investment Requests
  - Assists GM[a] in the strategy development and review
  - Assists GM in establishing and reviewing the 5-year business plan
  - Assists GM in the search and evaluation of acquisition partners
  - Assists GM in preparing and submitting Investment Requests for plant extensions, new plants, etc.
- Training
  - Covers the section "business development" in internal training programs

Competencies

- Establishment, together with GM of the
  - Yearly sales budget and
  - The yearly budget for his department
- Initiation of new R&D projects
- Conclusion of confidentiality agreements, R&D agreements, supply contracts (up to $1 million/year)
- business trips, customer entertainment (within yearly budget)
- participation at pertinent educational programs

Measures of Performance (MOP)

- Fulfilment of budget (EBIT, RONOA, ROI, etc.)
- Fulfilment of business plan (growth targets)
- Customer satisfaction (score card)
- Number of successfully established new business relationships
- Number and size of approved new product projects

Candidate Profile

- A paradoxical blend of personal humility and professional will[b]
- Advanced degree in organic chemistry
- Several years experience in a similar position (minimum annual sales: $10 million)
- Flair for commercial activities
- Familiarity with basic financial principles
- Staying power
- Perfect knowledge of English (written and spoken)

[a] General Manager.
[b] Jim Collins, *Good to Great*, 2001.

Checklist for the Selection of Outsourcing Partners

Criteria	Sub-criteria	Relevance
Innovation and technology	• Plant/facilities • Technology toolbox • Process development capability • Analytical development capability • Project management	****
Quality	• FDA record • Quality systems • Performance measures • Change control	****
Risk and security of supply	• Financial stability • Backup capacity • Approach to inventory • SHE approach • GRM audit • Political situation	***
Business attitude	• Secrecy • Reliability (on time/on budget) • Continuous improvement • Risk sharing • Capacity/lead time • Flexibility • Communication	****
Strategic fit	• Conformity with tax optimization policies • Conformity with market access strategies • Culture match	***
Price	• Total cost • Price reduction performance • Cost breakdown availability	**

Fine Chemicals: The Industry and the Business, Second Edition, by Peter Pollak
Copyright © 2011 John Wiley & Sons, Inc.

Checklist for the Manufacture of Nonregulated (*or* Basic GMP) Fine Chemicals

General aspects	Permits from authorities to run the plant
	Basic (simple) SOP-system
	Appropriate qualification and training program of the staff
	Documentation of master batch records and analytical methods process Documentation of process changes at communication to customer
Cleaning procedure	Limits for toxic compounds to be calculated on the basis of reactor volume
	Rinse with process solvents of the next production
	Visual check (to be documented)
Raw materials, intermediates, finished products; warehouse	Documentation of analytical methods, including changes
	Storage of documents for a minimum of 3 years
	Clear labeling for substances, piping, packaging, etc.
	Distinction between released and not released material on the packing (drums, barrels, etc.)
	Warehouse must safeguard product quality (humidity, temperature, dust, etc.)
Cross contamination	Design of workflow (organization plan) for avoiding cross contamination
	Suitable technical equipment
	Cleaning records to be kept for a minimum of 3 years
Maintenance/ Calibration	Maintenance program for the production unit and analytic department
	(Has to include a calibration program for production balances, thermometer. etc., and should be documented)
SHE guidelines	Plant operators must be properly trained
	Plant operators must wear suitable clothes (helmet, gloves, goggles, shoes), especially when working with toxic substances
	Availability and specifications of waste treatment plants
	Destination (documented) of all wastes (solid, liquid [aqueous/nonaqueous], gaseous)

Source: Boehringer Ingelheim.

Fine Chemicals: The Industry and the Business, Second Edition, by Peter Pollak
Copyright © 2011 John Wiley & Sons, Inc.

Checklist for Customer Visit

Timing	Action
−1 year	Prepare general plan for proactive visits for next year (A- and B-customers).
−1 month	Agree with customer on date, purpose, and participants.
−4 weeks	Prepare visit:
	(1) *Customer company information*:
	Study customer's website (especially annual report and news releases).
	Study customer's file and recent e-mail exchange.
	Check sales statistics.
	Ask other departments (R&D, purchasing, finance, legal, etc.) for pending issues.
	(2) *Product information*:
	Current business: check product files, especially sales statistics for individual products and active contracts.
	New business: check customer's information on development products, vertiefen durch consultation of databanks (e.g. SciFinder, www.cas.org/ SCIFINDER; Pioneer, www.ims-global.com), patent research.
	(3) *Agenda*:
	Prepare agenda.
	Depending on agenda, ask other departments to delegate specialist (e.g., local representative, R&D, production, legal, etc.).
−2 weeks	Confirm date and place of visit with customer and send agenda briefing with participating colleagues, define main objectives, prepare a list of adaptation of standard customer presentation (information you are looking for).
	Prepare gifts.
	When visiting key accounts, study files of participants from customers.

TABLE (*Continued*)

Timing	Action
On the road 0	Assign roles within participants of your company, rehearse, read newspaper of customer's country. Create a pleasant atmosphere, allow customer time to familiarize with agenda, present participants, say something positive on the customer. Give presentation. Discuss specific topics and try to accomplish preset goals. Summarize results of discussions and agree on next steps.
+1 day	Confirm "next steps" with customer in writing. Discuss results (1) with other participants: objectives met? who does what? what can be done better next time? (2) with other interested parties in your company. Write visit report and bring files up-to-date.
+1 month	Check execution of agreed-upon actions.

Outline for a Company Presentation

Slide 1	Title page (names of participants from **A** and **X** and agenda)
Slide 2	Company **A** at a glance: ownership, sales (a % standard products, b % exclusives), employees (manufacturing, R&D, sales and administration, etc.), major locations
Slide 3	Mission statement: "We want to be. ... Aspiration 2010: turnover $ ... million/EBIT $ ... million
Slide 4	Company history • History of **A** (foundation, major expansions, ownership changes) • History of the **A/X** relationship (first contact, first product supply, development of sales, current business)
Slide 5	**A** figures: sales development, geographical distribution, product categories, personnel development and distribution, investments)
Slide 6	**A** facility overview I: site plan, manufacturing buildings, production trains, reactor volume
Slide 7	**A** facility overview II: R&D, QC, and pilot plant (general view and key equipment)
Slide 8	**A** technological expertise: name reactions with volume ranges and examples of processes done on industrial scale
Slide 9	**A** SHE policy; and HSE: treatment of solid, liquid (aqueous and organic), and gaseous waste. Responsible care, corporate governance
Slide 10	Organization of **A** (management team, key contacts)

A, supplier; **X**, customer.

Overseas Expansion of Indian Pharma and Fine Chemical Companies

Acquirer	Acquired Company	Activity	Purchase Price
Avesta Biotherapeutics & Research	Siegfried Biologics (Germany)	Biologics development	N/A (50 employees)
Dishman Pharmaceuticals & Chemicals	Carbogen & Amcis[a] (Bubendorf, Switzerland)	CRO and 100-kg scale	$74.5 million (1.1 × sales)
	Synprotec (Manchester, UK)	CRO	$ 3.8 million
	Solvay Pharmaceuticals Fine Chemicals and Vitamin Business (The Netherlands)	Vitamin D and analogs	N/A ("much less than Carbogen-Amcis")
Dr. Reddy's Laboratories	Trigenesis (USA)	Dermatologics	$11 million
	Roche's API plant in Cuernavaca (Mexico)	Naproxen, steroids	$59 million
	Dowpharma's UK plants (Mirfiled and Cambridge)	Small molecules CM	$25 million
Glenmark Pharmaceuticals	Instituto Biochimico Industriale (Brazil)	CRO	$4.6 million
HIKAL	Marsing (Copenhague, Denmark)	API trader	$6 million (50.1% stake)

Fine Chemicals: The Industry and the Business, Second Edition, by Peter Pollak
Copyright © 2011 John Wiley & Sons, Inc.

TABLE   *(Continued)*

Acquirer	Acquired Company	Activity	Purchase Price
Jubilant Organosys	Cambrex (USA)	Fine Chemicals	$500 million
	Target Research Associates (USA)	CRO	$33 million
Malladi Drugs & Pharmaceuticals	Novus Fine Chemicals (Carlstadt, NJ, USA)	Ephedrine, pseudo-ephedrine	$23 million
Matrix Laboratories	Explora Laboratories (Mendrisio TI, Switzerland)	CRO	$20 million (50% stake)
	Fine Chemicals (South Africa)	APIs	$ 263 million
Nicholas Piramal India (NPIL) Ltd.	Avecia[b] (Huddersfield, UK)	PFC (CRO)	≈€25 million (sales €58 million)
NPIL (UK) Ltd	Pfizer's manufacturing site (Morpeth, UK)	APIs and finished dosage forms	$38 million
	Torcan (Canada)	CRO	N/A
Shasun Chemicals & Drugs	Rhodia Pharma Solutions	CM	N/A
Sun Pharmaceutical	ICN (Hungary)	Morphine, codeine	N/A
Suven Life Sciences	Synthon Chiragenics (Monmouth Junction, NJ)	Carbohydrate-based chiral technology	N/A

[a] Solutia Pharmaceutical Services Division.
[b] Including Torcan CRO, Canada.

Asian Expansion of Western Fine Chemical Companies

Western Investor	Asian Venture	Activity	Characteristics
AMRI	Acquiring of two PFC plants (India)	Fine chemicals	$11 million (2007); + $15 million (2010) (2008–2011)
Aptuit (USA)	JV with Laurus Labs (India)	CRO	$100 million (2007–2010)
DSM (NL)	Manufacturing Agreement with Arch Pharmalabs (India)	Fine chemicals API-for-Generics	(2007)
	Construction of corporate R&D laboratories Zhangjiang Hi-Tech Park (China)	R&D	(2008)
Excelsyn (UK)	Shanghai Organic Synthesis Development (OSD) (China)	Fine chemicals	"Multimillion $" (2004)
Evonik (Germany)	JV with Lynchem (Dalian, China)	Fine chemicals	(2008)
Flamma (Italy)	Dalian Biosciences (China)	CRO	(2008)
Lonza (Switzerland)	API and intermediates (Guangzhou, China)	Fine chemicals	$200 million investment
Nycomed (Switzerland)	JV Zydus Nicomed (India)	API manufacturing	1999[a] ($50 million)

TABLE   (*Continued*)

Western Investor	Asian Venture	Activity	Characteristics
Merger between PacificGMP, USA and Pacific Biopharma Group, Cayman, to form China Quantitative Biomedecine (Shanghai)		Therapeutic proteins by "single use disposable technology"	N/A
Pfizer (USA)	Manufacturing agreement with HIKAL (India)	CM	(2008)
	Licensing agreement with Biocon (India)	Marketing of insulin products	2010
Pfizer CentreSource (USA)	Shanghai Pharmaceutical Group (China) ScinoPharm, Taiwan	Fine chemicals (chemical steps of steroid synthesis)	2006 2008
SAFC (USA)	Medicinal chemistry facility Bangalore, India	CRO	$12 million investment (2006)

ª JV with Byk Gulden.

Printed in the USA/Agawam, MA
April 16, 2012
565368.014